3/06

D0843240

# Volcanic Worlds
Exploring The Solar System's Volcanoes

**Springer**
*Berlin*
*Heidelberg*
*New York*
*Hong Kong*
*London*
*Milan*
*Paris*
*Tokyo*

Rosaly M. C. Lopes and Tracy K. P. Gregg

# Volcanic Worlds

## Exploring The Solar System's Volcanoes

Springer

Published in association with
**Praxis Publishing**
Chichester, UK

Dr Rosaly M. C. Lopes
Jet Propulsion Laboratory/NASA
Pasadena
California
USA

Dr Tracy K. P. Gregg
The State University of New York at Buffalo
Buffalo
New York
USA

SPRINGER–PRAXIS BOOKS IN GEOPHYSICAL SCIENCES
SUBJECT *ADVISORY EDITOR*: Dr Philippe Blondel, C.Geol., F.G.S., Ph.D., M.Sc., Senior Scientist, Department of Physics, University of Bath, Bath, UK

ISBN 3-540-00431-9 Springer-Verlag Berlin Heidelberg New York

Springer-Verlag is a part of Springer Science+Business Media (springeronline.com)

Bibliographic information published by Die Deutsche Bibliothek

Die Deutsche Bibliothek lists this publication in the Deutsche Nationalbibliografie; detailed bibliographic data are available from the Internet at http://dnb.ddb.de

Library of Congress Control Number: 2004105865

Cover design: Jim Wilkie
Project Management: Originator Publishing Services, Gt Yarmouth, Norfolk, UK

Printed on acid-free paper

# Contents

# Foreword

When I made my first voyage into space 20 years ago, the book you are holding would not have been possible. Over the last two decades, our understanding of volcanism and its importance has changed significantly—and so have the faces of those who study it.

*Volcanic Worlds* is the first edited book on planetary geology (and to my knowledge on any topic in astronomy) written solely by women. This has only become possible recently, and clearly illustrates the changing landscape of science.

Today, women contribute to our knowledge in every scientific field. They are engaged in exploring some of our most fundamental questions, including the subject of this book: the role that volcanoes play in our solar system, and what the internal churnings of planets and moons tell us about their origin and their future.

All the authors are respected experts in planetary science. They will take you on a fascinating tour of the solar system's richly diverse volcanoes and volcanic processes. You'll explore the tumultuous young Earth, and travel from a suffocatingly hot Venus to the icy moons of the outer solar system.

You will also meet the intriguing women behind each chapter. You ll learn about their career paths and research in their own words. Among them are a JPL scientist responsible for the discovery of more than 70 volcanoes, an astrogeologist engaged in ongoing discussions for a return voyage to the Moon, and a professor who has ventured to every continent to study Earth's volcanoes.

These women have followed different paths into science, but their careers share a spirit of perseverance, ambition, independence, and occasionally, a touch of serendipity. As you read about them and their research, you'll see there's no doubt that the presence of women in science is secure—and will continue to grow in future generations.

In 1847, a young New Englander became the first American to record seeing a comet through a telescope. Maria Mitchell looked through her telescope and noticed a star a few degrees from the North star that she didn't think had been there before. The next night, the "star" had moved and she knew the blurry image was a comet. She went on to become the first female astronomy professor in the United States, and developed a reputation for holding her Vassar College students to the highest standards despite their being "just women".

Maria Mitchell had the right idea. Times have certainly changed—and science is benefiting because of it.

*Sally Ride*
Professor of Physics, University of California San Diego
President, Sally Ride Science
April 2004

# Acknowledgements

We are greatly indebted to the editors and staff at Praxis and Originator for their support, patience, and ideas on how to improve our manuscript. In particular we would like to thank Clive Horwood, Philippe Blondel, John Mason, and Neil Shuttlewood. Thanks are also due to Jim Wilke for the cover design and to Catherine Ivy for bringing our manuscript to the attention of Sally Ride. We are all very grateful to family and friends for their support and forebearance. We would also like to thank numerous male colleagues for their encouragment and for recognizing that this book illustrates the remarkable progress that women have made in science during recent times.

# Figures

# Tables

# Color plates (between pages 112–113)

# 1

# Volcanic Worlds: Introduction

## *Tracy K.P. Gregg and Rosaly M. Lopes*

Volcanism is a fundamental process that has affected every solid body throughout our solar system and, presumably, in solar systems beyond ours. As we explore other worlds, we come across forms of volcanism much different from those we see on Earth today, and others that are surprisingly similar. By understanding volcanoes on other planets, we can determine how the Earth may have behaved under diverse conditions that may have existed at different times during Earth's history—or its future. If a person's eyes can be seen as windows into their soul, then a volcano can be thought of as a window into a planet's soul. Until we are able to drill to the center of the Earth, or any other planet, volcanoes offer us a glimpse into the deep interior workings of a planetary body, showing how they may have changed with time.

This book is intended as a presentation of the diversity of manifestations of planetary volcanism, and what this tells us about the diversity of planets, their formation, and their evolution. We can truly learn a great deal from volcanoes. But what are they exactly? Strictly defined, a volcano is a place on the solid surface of a planet where relatively hot material from the planet's interior comes out. On Earth, we normally think of volcanoes as molten rock coming to the surface (as explained in Chapter 2). But a geyser—where hot water comes from the Earth's interior—is also a form of volcano by this definition (Chapter 11). This definition allows us to go to the outer solar system, where a planetary surface may be composed of frozen water or methane, and the erupting "lava" is actually molten ice—also known as water (Chapter 9).

Volcanism is merely a mechanism by which a planet-sized body (a term that includes solid objects in the solar system that are around the same size as our moon, or larger) loses its internal heat. At some point during solar system formation, all planetary bodies received heat from one or more processes. Debate still exists about precisely when these events may have taken place, although most scientists agree that they occurred largely within the first billion years of the existence of the solar system. Furthermore, not all of these events necessarily provided heat to all the planetary bodies in the solar system. Regardless, sources of heat for the early planetary bodies were:

1   Heat of accretion. The early solar system was filled with large chunks of rocks (hundreds of kilometers across) frequently impacting upon the forming planets. The kinetic energy from the impacts would then be converted into thermal energy. It is believed, for example, that one of these large pieces of space debris (the size of Mars) smashed into early Earth. This collision resulted in the formation of our moon (see Chapter 5).

2   Core formation. If a planetary body received enough thermal energy (through accretion), the planet could have *differentiated*. This means that the planet was sufficiently molten so that the heavy elements, like iron, could sink toward the center of the planet. This certainly happened on Earth, and probably on all other planetary bodies as well (although to varying degrees). This process would have released heat in two ways. First, the sinking, dense material would rub against the stationary material, generating friction and heat. Second, the sinking process itself would convert some of the potential (gravitational) energy into thermal energy.

3   Radioactive isotopes. The primary source of heat on the Earth today is through the decay of naturally occurring radioactive materials in the planet's interior. Some radioactive materials decay more quickly than others, and there were many fast-decaying materials early in the solar system's history that were a significant source of heat—until they literally decayed away.

4   Tidal heating. This is essentially friction. A planetary body may be gravitationally attracted by other planets, depending on the respective orbits of these bodies. Every time a planet is tugged, friction is created in its interior, because not all parts of the planet are tugged at the same rate or with the same force. Friction creates heat, as anyone who has rubbed their hands together to warm them knows. Io is the prime example of a planetary body heated by tidal forces in our solar system (Chapter 8).

5   Solar heating. On most solid planets, the Sun's energy is too weak to drive volcanic activity—although on Neptune's moon Triton, geysers erupt every summer, driven by solar power (Chapter 11).

Nature abhors differences. Erosion will level mountains, ocean basins will fill with sediments, and all planets want to be the same temperature as outer space. Each planetary body has a unique way of attempting to reach this equilibrium, using volcanism.

Earth, for example, is the only planet in the solar system to display plate tectonics (Chapters 2 and 3). The Earth's outer layer is not stable, and is broken up into a series of pieces moving across the globe. Volcanoes tend to be concentrated where these plates come together and move apart (see Chapter 3). Io might be considered another extreme: a planet where internal heat is released through huge volcanoes that dot the surface in a seemingly random pattern (Chapter 8). Volcanoes on Venus are neither randomly distributed nor associated with plate boundaries

(Chapter 4). Instead, volcanoes on Venus appear to be associated with faults and fractures, although they are not arranged in a global pattern. Mars has the largest volcano in the solar system, called Olympus Mons (Chapter 6); it is five times taller than Mount Everest. In spite of its size, it looks very much like a shield volcano on Earth (Chapter 2), although some scientists believe that ice may have played an important role in its formation (Chapter 10). There are volcanoes on Mars that are unlike any we have seen on any other planet (Chapters 6 and 10). As we write these lines, several of us are actively involved in the analyses of data from the 2004 missions on the Martian surface. This means that our interpretations of that planet are literally changing every hour. The current thinking, as well as the techniques used are presented in Chapter 7.

Prior to the Voyager missions of the 1970s and 1980s, Earth was thought to be the only planet in the solar system with active volcanism. The Apollo lunar missions revealed that the Moon has not experienced any active volcanism for millions (possibly billions) of years (Chapter 5). Viking missions had not revealed any activity on Mars (Chapter 7), and the available Venusian data were not terribly suggestive of activity (although the Pioneer Venus spacecraft did detect a sudden increase—followed by a decrease—in the amount of sulfur in the atmosphere while orbiting Venus; this may have been a volcanic emission). Any planet smaller than Mars, it was reasoned, should have frozen to death hundreds of millions of years ago or more.

Since then, we have had to change our ideas of what "active" means. The Voyager I spacecraft imaged an erupting volcano on Io; since then, at any given time, Io appears to have dozens of volcanoes erupting on it (see Chapter 8). The surface of Europa looks so fresh and young that it is probably less than 30 million years old; and only recent volcanic eruptions (here, in the form of ice volcanism; Chapter 9) could be responsible. Geysers were observed erupting on Neptune's moon Triton, a planet smaller than our own moon, and are likely to be driven by solar power (Chapter 11). Increasingly high-resolution images of the Martian surface have also made some geologists rethink what they mean by the term "active", since lava flows younger than a million years have been discovered there (see Chapter 6).

Volcanoes have erupted on almost every solid planetary surface in the solar system. These planets range from the small but energetic Triton in the outer reaches of the solar system, to Mercury, nestled close to the Sun and still waiting to reveal almost 70% of its surface to our instruments (Chapter 5).

Rather than stick to a traditional presentation (i.e., chapter by chapter), for the contents of this book, we have instead tried to relate them to the scientific discoveries of only the last decades. In any order, however, the following chapters reveal the immense diversity of planetary volcanism. We, the women who study these volcanoes and wrote this book, are equally diverse, but all of us share the same thrill when we discover something new about the workings of these fascinating and mysterious places. We hope to pass on this enthusiasm (and all this fresh knowledge) to our readers.

# 2

# Volcanoes on Earth: Our basis for understanding volcanism

*Katharine V. Cashman* (University of Oregon)

*I did not always know that I wanted to be a scientist. As a child, my combined passions were books and the outdoors, and even when I entered college I was torn between a major in English and a major in natural science. However, two summers as a geology field assistant in the Klamath Mountains, together with the inspirational teaching of Dr. Peter Coney, soon led me down the path of geology. Having learned geology in New England, I did not become acquainted with active volcanoes until after college, when I received a Fulbright Scholarship to study in New Zealand. There I learned to love volcanic landscapes, and from there I had the chance to travel south to the frozen continent of Antarctica, where I first worked on an active volcano—Mount Erebus. Even then I did not commit to volcanology, instead returning to the U.S.A. to take a job with the US Geological Survey (USGS) as a marine geophysicist. My career as a volcanologist took shape in 1980, with the eruption of Mount St. Helens. Unable to bear marine studies on the East Coast when a volcano was erupting in Washington state, I begged for a transfer and accepted the position of Public Information Scientist for the USGS Cascade Volcano Observatory. One and a half years of talking to the media was sufficient encouragement for me to return to school for my PhD, which I obtained at the Johns Hopkins University in Baltimore, Maryland. I then taught for five years at Princeton University before transferring to the University of Oregon, where I am now. A career in volcanology has allowed me unprecedented opportunity for travel—I have visited volcanoes on all seven continents, and have volcanologist friends around the world. I never tire of watching active lava flows in Hawaii, or of climbing volcanoes in Alaska or Ecuador, Oregon or New Zealand, Italy or Japan. Nor do I tire of studying the products of volcanic eruptions, particularly through the use of an electron microscope, which allows me to enter the magical world of the microscopic. Ultimately, however, my research is driven by the puzzles that volcanoes provide, the mysteries of how they work, the majesty of the landforms that they construct, and the hope that something I do may someday help people to live more safely in their shadows.*

**Figure 2.1.** Picture of the author in front of the Erebus summit hut in 1978 (accompanied by mascot Paddington).

## 2.1  INTRODUCTION

*Keeper of the Southern Gateway, grim, rugged, gloomy and grand*
*Warden of these wastes uncharted, as the years swept on, you stand*
*At your head the swinging smoke cloud; at your feet the grinding floes;*
*Racked and seared by inner fires, gripped close by the outer snows.*[1]

The wind whipped my parka, nestling the fur trim against my cheek. I took a deep breath and leaned over the edge of the crater, allowing the radiated warmth from the lava lake below to take away the chill. The glowing lava moved slowly, convecting, cooling, and blackening as molten rock met cold Antarctic air. I was at an elevation of 13,000 ft, on the summit of the volcano Mount Erebus, Antarctica, in December of 1978, and I was seeing hot lava for the first time in my life. In retrospect I think that moment decided my career (Figure 2.1).

An active volcano in Antarctica, the frozen continent, often strikes people as incongruous. Even its name—Erebus, the son of Chaos in Greek mythology, the shadowy underworld en route to Hades—is oddly classical in this uninhabited

[1] From the poem *Erebus* published in E.H. Shackleton (1988) *Aurora Australis*. SeTo Publishing, Auckland.

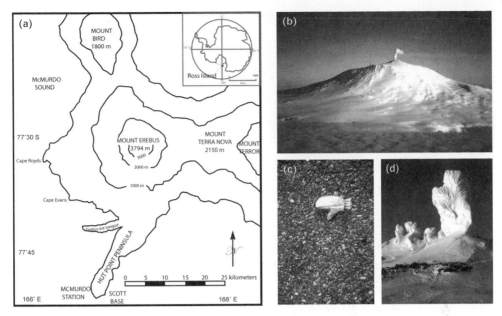

**Figure 2.2.** (a) Map of Ross Island, Antarctica, showing the location of the active volcano Mount Erebus, older volcanoes Mounts Bird, Terra Nova, and Terror, and locations mentioned in the text. (b) Photo of Mount Erebus showing a characteristic steam plume from the summit vent. (c) Scree of anorthoclase feldspar crystals on the slopes of the Erebus summit cone; mitten for scale. (d) Ice towers formed over steam vents (fumaroles) on the slopes of Erebus.

region. Mount Erebus was named for one of the two ships used by Capt. James Clark Ross on his 1840 Antarctic expedition. It forms the center of Ross Island (Figure 2.2(a)), which is connected to the Antarctic continent by the Ross Ice Shelf, a vast ice sheet that covers the southernmost 500,000 km$^2$ of the Ross Sea. The ship *Erebus* later met its fate at the opposite end of the globe as one of the vessels used by the ill-fated Franklin expedition, lost in the Arctic trying to find the Northwest Passage. Ross Island served as the winter headquarters for Capt. Robert Scott (1901–1904 at Hut Point and 1911–1913 at Cape Evans) and Sir Ernest Shackleton (1907–1909 at Cape Royds) in their early exploration of the Antarctic continent. Shackleton's men were the first to ascend the volcano and to describe periodic explosions from the summit crater, large feldspar crystals that comprise its summit cone, and intricate ice sculptures formed over steam-emitting fumaroles by contact with the cold Antarctic air (Figure 2.2(b–d)). Erebus expeditions now work out of modern bases (such as the US McMurdo Station and New Zealand Scott Base), and use helicopters, GPS navigational systems, and high-tech volcano monitoring tools—a far cry from a century ago. The motivation for study, however, has changed little over the past century. Volcanologists are still driven by the adventure of remote environments and the quest to understand how volcanoes work.

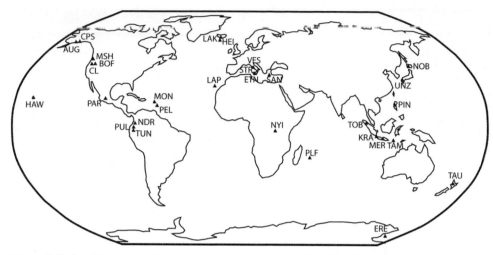

**Figure 2.3.** Location map showing volcanoes mentioned in this chapter. Abbreviations are as follows: HAW—Hawaii; AUG—Augustine, USA; CPS—Crater Peak, Mt. Spurr, USA; MSH—Mount St. Helens, USA; BOF—Big Obsidian Flow, USA; CL—Crater Lake, USA; PAR—Parícutin, Mexico; MON—Soufriere Hills, Montserrat; PEL—Mt. Pelee, Martinique; NDR—Nevado del Ruiz, Columbia; PUL—Pululahua, Ecuador; TUN—Tungurahua, Ecuador; LAK—Laki, Iceland; HEI—Heimaey, Iceland; LAP—La Palma, Canaries; VES—Vesuvius, Italy; STR—Stromboli, Italy; ETN—Mt. Etna, Italy; SAN—Santorini, Greece; NYI—Nyiragongo, Zaire; PLF—Piton de la Fournaise, Réunion; TOB—Toba, Indonesia; KRA—Krakatau, Indonesia; MER—Merapi, Indonesia; TAM—Tambora, Indonesia; UNZ—Unzen, Japan; NOR—Noboribetsu, Japan; PIN—Pinatubo, Philippines; TAU—Taupo, New Zealand.

Scientists who study terrestrial volcanism have an additional motivation for their work—that of attempting to mitigate the impact of volcanic eruptions on human populations. Volcanic regions are often heavily populated, as humans benefit from the fertility of volcanic soils. At the same time, the deleterious effects of volcanic eruptions have long been recognized, handed down through the ages via the spoken word, written records, and evidence of buried cities. Catastrophic eruptions, such as the 1883 eruption of Krakatau, Indonesia, and the 1902 eruption of Mount Pelee, Martinique (Figure 2.3) spawned the new discipline of volcanology, as a direct response to the devastating loss of life in these two events. Eruptions over the past few decades—Kilauea volcano, Hawaii (1983–present); Mount St. Helens, Washington (1980–1986); Unzen, Japan (1991–1995); Pinatubo, Philippines (1991); Soufriere Hills volcano, Montserrat (1995–present)—have propelled the field forward, giving rise to new monitoring tools and new interpretations of eruptive processes. At the same time, disastrous eruptions such as that of Nevado del Ruiz, Columbia (1985) have promoted a new awareness of the growing vulnerability of human populations to volcanic events. Throughout this chapter I attempt to address both the human and the scientific perspectives of volcanic activity by interspersing eyewitness accounts of, and human responses to,

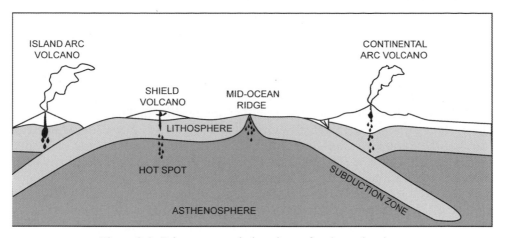

**Figure 2.4.** Primary tectonic locations of active volcanism.

historic eruptions with descriptions of volcanic landforms and processes. In each section, I provide an overview of the type of activity being discussed, followed by a description of the morphology of volcanic features, the processes involved in different types of eruptions, and the physical products of each event. Finally, I summarize the types of volcanic hazards presented by each volcanic environment. As our ability to monitor and document processes occurring at terrestrial volcanoes far exceeds our ability to study volcanoes on other planets, these descriptions of terrestrial volcanism also serve as an underpinning for interpretations of volcanic activity on other planets.

## 2.2   HOT SPOT VOLCANISM

> *The activity of the crater is violent and rough*
> *It bursts on high, breaking pointedly by the storm . . .*
> *The stones grow in agony*
> *From the clawing of the fire . . .*
> *The island is drawn up and flattened down*
> *The heavens are low and the mountain is surging*
> *The ocean dances, Kilauea surges . . .*[2]

The volcano Erebus lies above a mantle plume, or "hot spot", so-named because it marks the location of persistent melt production in the Earth's mantle (Figure 2.4). Here melt is generated by rise (decompression) of the asthenosphere, perhaps from as deep as the boundary between the Earth's mantle and core. Partial melting of the mantle creates a basaltic liquid that is composed of about 50% $SiO_2$

[2] P.K. Kanahele (2001) *Holo Mai Pele*. Native Books, Inc., Honolulu, 67 pp.

and 50% oxides (oxygen compounds) of Al, Fe, Mg, Ca, Na, K, and Ti. This magma (liquid silicate melt +/− crystals and dissolved gases) erupts at the Earth's surface as lava (liquid silicate melt +/− crystals) to form volcanoes.

Perhaps the most famous hot spot on Earth is that responsible for the volcanoes of Hawaii, which lie at the south-eastern end of an extensive chain of volcanic islands—the Hawaiian-Emperor chain—that stretches across much of the north Pacific basin. Here a hot spot has persisted for tens of millions of years. As a result, the ages of islands created by the hot spot march backwards through time, marking the north-westward movement of the Pacific plate over the fixed hot spot source (Figure 2.5). The Hawaiian islands themselves represent a microcosm of the longer Hawaiian-Emperor chain, with island ages decreasing from north-western Niihau (~5 million years in age) to south-eastern Hawaii, the current location of the hot spot. The island of Hawaii shows the same progression in miniature, with its five separate volcanoes—Kohala, Mauna Kea, Hualalai, Mauna Loa and Kilauea—also decreasing in age (from ~430,000 years to the present-day) and increasing in activity to the south-east. The newest Hawaiian volcano—Loihi—lies offshore of Kilauea, one kilometer beneath the ocean's surface, and marks the end of the chain.

The south-eastern progression of active volcanism is described by Hawaiians in the story of Pele, the deity responsible for all manifestations of volcanic activity in Hawaii. According to Hawaiian mythology, Pele and her extensive family traveled from Tahiti to a point north-west of Hawaii. As Pele traveled southeast down the island chain she dug successive fire pits in search of a suitable home, always chased by her sister the ocean, Na-maka-o-ka-hai, until she settled on Kilauea volcano, thus tracing the hot spot track. This story shows that the Hawaiians understood the relative age of the islands, their mode of formation, and the importance of lava–water interactions in constructing coastal (littoral) cones such as Oahu's famous Diamond Head. Inferring geologic processes based on their resemblance to current activity is an important geologic concept, first articulated by the famous British geologist Charles Lyell in 1830. This principle is known as *uniformitarianism*, a concept summarized as "the present is the key to the past", and of fundamental importance to studies of both terrestrial and planetary volcanism.

### 2.2.1    Volcano morphology

The basaltic volcanoes of Hawaii are classic examples of shield volcanoes, named for their broad convex forms. They are built primarily of thin lava flows and may be very large—Mauna Loa volcano, at 40,000 km$^3$, rises 4,169 m above sea level and 8 km above the surrounding ocean floor (Figure 2.6(a)). Young shield volcanoes typically have summit calderas (circular or elliptical depressions) flanked by elongate rift zones through which lava is transported away from summit reservoirs. Resulting flank eruptions emerge from linear fissure vents that send broad sheets of lava downslope (Figure 2.6(b)). Such eruptions create low spatter ramparts that may

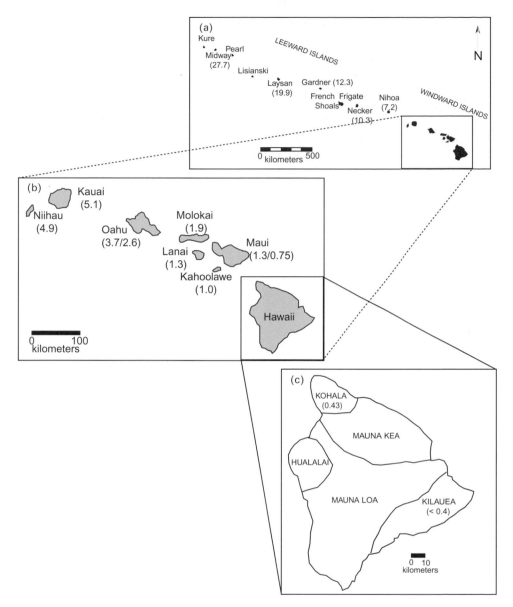

**Figure 2.5.** Maps illustrating the formation of the Hawaiian island chain; volcano ages are given in brackets in millions of years. (a) Leeward and windward islands of the Hawaiian chain. (b) Close-up of the Hawaiian islands. (c) Close-up of the island of Hawaii, where individual volcanoes are delineated.
Redrafted from USGS maps.

**Figure 2.6.** Photographs illustrating morphological features of Hawaiian volcanoes. (a) Mauna Loa, a classic shield volcano. (b) Fissure eruption along Kilauea's north-east rift zone. (c) Late stage cinder cones on Mauna Kea's summit. (d) Mauna Kea, illustrating cinder cone coverage on top of an older shield.
(b) USGS photograph.

focus into a single vent, where continued eruptive activity may create cones of scoria, or cinder (frozen bubbly basaltic melt; Figure 2.6(c)). In later stages of their evolution, the rift zones of Hawaiian volcanoes cease to function as magma conduits, and shield surfaces become dotted with small cinder cone vents (Figure 2.6(d)).

The low slopes of shield volcanoes are produced by accumulation of thin, fluid (low viscosity), basaltic lava flows. Basaltic lava flows tend to be much longer than they are wide, and much wider than they are deep. The length of flows with simple emplacement histories is determined primarily by the eruption rate (Figure 2.7). The slope over which lava flows travel also influences flow length, and together with eruption rate controls both the surface roughness and the planform (map view) shape of individual flows. The combined effect of eruption rate and slope is best seen in the relative distribution of the two types of basaltic lava flows: smooth-surfaced pahoehoe and rough aa lava (Figure 2.8(a,b); described in more detail below). Large pahoehoe flow fields develop during protracted eruptions that deliver lava through well insulated lava tubes to near-horizontal coastal plains. In contrast, large aa flows are emplaced rapidly through lava channels formed on the steeper slopes along the rift zones. Thus Hawaiian volcanoes are constructed primarily by addition of pahoehoe in areas of low slopes and aa on steeper rift zone flanks (Figure 2.8c).

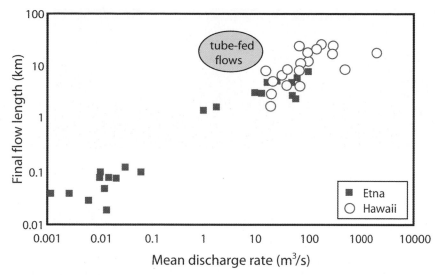

**Figure 2.7.** Plot showing the relationship between final flow length and eruption rate for simple channel-fed lava flows from Hawaii and Etna. Shown in the field in gray is the field of Hawaiian flows fed by lava tubes.
Modified from Kilburn (2000).

## 2.2.2    Fire fountains

Modern visitors to Hawaii hope to witness the spectacular lava fountains so closely associated with the island of Hawaii that this type of activity is termed *Hawaiian*. Fire fountains occur when magma travels through volcanic rift zones to a point where it ruptures the surface, creating a dancing display of individual lava fountains (a "curtain of fire") that may extend for kilometers downrift (Figure 2.6(b)). Multiple fountains soon focus into a single vent to create fire fountains that send lava shooting upward to heights of 100–500 m (Figure 2.9(a); see color section). Lava fountains are driven by the rapid formation and expansion of gas bubbles within the liquid lava, gas that is dissolved in the magma under pressure but is released (exsolved) at the low pressures of Earth's surface. Gas expansion accelerates the bubbly mixture until it breaks apart like the spray of a high-pressure garden hose to form large aerodynamically shaped lava fragments (bombs), bubbly clasts (scoria), small melt droplets (Pele's tears), and fine glass stringers (Pele's hair). High temperatures and rapid decompression of melt in the center of lava fountains allow gas bubbles to expand and connect (coalesce) before cooling to glass (quenching). Bubbles preserved in this quenched melt are often spherical, but in high fire fountains the bubbles may expand to form delicate polyhedral foams called reticulite (Figure 2.9(b)).

Fire fountains form when magma rises rapidly from the subsurface, and bubbles are contained within the melt as they expand. In contrast, slow ascent of magma allows bubbles to separate from the melt, rise, and coalesce to form large individual

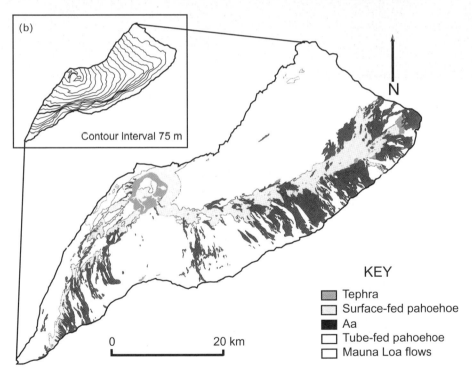

**Figure 2.8.** Surface morphology of Kilauea volcano. (a) Typical pahoehoe and lava flow surfaces. (b) Map showing the distribution of flow types on Kilauea volcano (inset shows topography of the same area).
Modified from Holcomb (1987).

**Figure 2.9.** Photographs illustrating the products of Hawaiian fire fountains. (a) Fountaining episode of Kilauea volcano, July, 1984. (b) Reticulite (photograph ∼3 cm across).
(a) See color section also.
(b) USGS photograph by M. Mangan.

bubbles up to meters in diameter. These large bubbles rise through the magma-filled conduit to rupture at volcanic vents, producing a type of eruptive activity named *Strombolian* for the Italian volcano Stromboli in the Aeolian Islands north of Sicily (Figure 2.10(a)). Stromboli's typically mild and persistent eruptive activity produces a bubble-bursting event every 10–20 minutes, and has done so at least since the time of Homer (8th century BC). However, larger explosions occur when deep, gas-rich magma rises rapidly to the surface instead of stalling at shallow levels. Such an event occurred on 5 April, 2003, and sent large blocks into the town of Ginostra on the west coast of the island (Figure 2.10(b,c)). Persistent, although less regular, Strombolian eruptions occur at other basaltic volcanoes around the world, including Mount Etna (Sicily, Italy) and Mount Erebus (Antarctica).

### 2.2.3  Lava flows

Most commonly visitors to Hawaii do not have the opportunity to view fire fountains, but instead experience Pele more intimately by visiting coastal lava

**Figure 2.10.** (a) Stromboli volcano. (b) Volcanic bomb damage to a house in the village of Ginostra, 5 April, 2003. (c) Volcanic "shrapnel" from impact breakage of a volcanic bomb from the same event—bomb fragment is embedded in a prickly pear cactus.

flows that have dominated Kilauea's activity since 1986. Frequent lava flows, however, are neither restricted to Kilauea's current eruption nor unique to Kilauea volcano. When Rev. William Ellis visited the island of Hawaii in 1823 he was told that Kilauea had been "burning since time immemorial, and...that eruptions from it had taken place during every king's reign...from Akea, the king of the island, down to the present monarch".[3] An eruption of Hualalai volcano in 1800–1801 also impressed an early European inhabitant of Hawaii, who told Ellis that he was "astonished at the irresistible impetuosity of the torrent", an apt description of the awe produced by large swiftly flowing channels of molten lava. This eruption, and one that probably occurred about 50 years earlier, are the most recent manifestations of volcanic activity on Hualalai volcano, whose slopes now host the tourist destination of Kona. Kilauea's neighbor, Mauna Loa, has produced very large lava flows, particularly in the late 19th and early 20th centuries, when lava flows occurred at a rate of one every 3–5 years. Although eruptions of Mauna Loa have been less frequent over the past decades, an energetic eruption in 1950 sent lava flows pouring down the steep slopes of Mauna Loa's south-west rift zone, while more recently, lava flows from Mauna Loa traveled to within 7–8 km (5 miles) of the outskirts of the town of Hilo in 1984.

Both the most recent eruption of Mauna Loa in 1984 and the ongoing eruption of Kilauea have provided extensive opportunities for studying the transport and emplacement of lava flows, and for quantifying aspects of Hawaiian eruptions described by early observers. Yale University's Prof. J.D. Dana first introduced the Hawaiian words *pahoehoe* and *aa* to the scientific community in 1849,[4] and was one of a long list of scientists to be fascinated by the origin of these two different flow types. Early workers recognized that both pahoehoe and aa could be produced from the same lava during the same eruption, thus concluding that the difference lay in the physical conditions of flow emplacement rather than in the chemical composition of the basalt. Recent work has shown that complex feedback between rates of flow cooling, crystallization, and flow advance, are required to explain the transition of lava from one form to the other.

High intensity eruptions produce large lava channels with surfaces that may remain partially free of solid crust for many kilometers. W.D. Alexander described such an eruption of the Mauna Loa volcano in 1859 as, "a cataract of fire (which) continued for several miles in a winding river of light, which then divided into a network of branches, enclosing numerous islands."[5] These islands are now known by their Hawaiian name, *kipuka* (Figure 2.11(a); see color section also). As flows travel away from eruptive vents, the channel surface cools and solidifies, creating a dark stripe of solidified lava down the center of the flow. High shear rates along channel margins and flow fronts, however, tear the fragile surface crust (Figure 2.11(b); see color section also), thus exposing hot interior lava to surface cooling. Cooling causes small crystals to form within the liquid (Figure 2.11(c)). When the crystal content is sufficiently high (20–30%), these crystals interact during flow to generate rough aa clinkers. Solidified clinkers on the flow

[3] W. Ellis (1963) *Journal of William Ellis—narrative of a tour of Hawaii, or Owhyee; with remarks on the history, traditions, manners, customs and language of the inhabitants of the Sandwich Islands.* Advertiser Publishing company, Ltd., Honolulu.
[4] J.D. Dana (1849) *Geology: United States Exploring Expedition.* Putnam, New York.
[5] W.D. Alexander (1859) Later details from the volcano on Hawaii. *Pacific Commercial Advertiser*, p. 2.

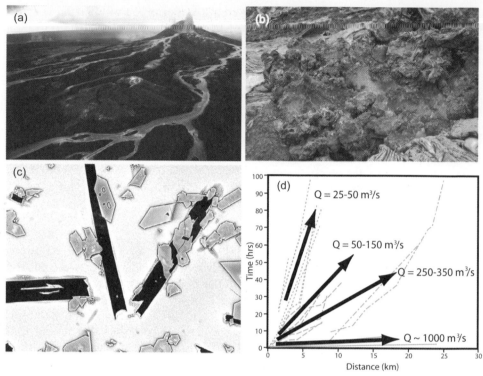

**Figure 2.11.** (a) Open channel lava flows on Kilauea volcano. (b) Close-up photograph of aa flow front. (c) Scanning electron microscope image of aa samples showing tiny crystals of plagioclase (black) and pyroxene (gray) that grew within the melt (now glass, light gray) during flow through an open channel (image is 125 μm across). (d) Changes in lava flow advance rate with variations in eruption rate $Q$ (shown in m$^3$/s)—higher eruption rates result in faster flows.
Modified from Kauahikaua *et al.* (2003).
(a) USGS photograph by J. Griggs. (a) and (b) see color section also.

surface move outward to flow margins as flows spread, forming levees. These levees construct a channel that aids flow of lava to the flow front.

Comparison of recent flows from Kilauea and Mauna Loa volcanoes shows that initial rates of flow advance are controlled primarily by the eruption rate (Figure 2.11(d)). The fastest flows of recent history are those of Mauna Loa's 1950 eruption, which covered the 24 km (15 miles) from the vent to the ocean in less than three hours. Typically, lava flows in Hawaii advance at rates of <0.1 to ~10 km/hour. In contrast, lava that is later transported through the established channel may travel much more rapidly, at speeds up to about 50 km/hr. However, not all lava flows produce open channels. If flow advance is sufficiently slow, rapidly formed surface crust will encase the flow in an insulating cover of solidified lava, thus forming a lava tube. Lava tubes can transport lava over great distances with very little cooling (Figure 2.12). Typical cooling rates are <1°C/km, which translates to ~1–2°C/hr

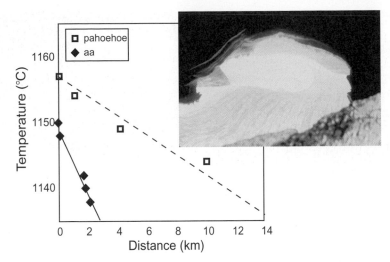

**Figure 2.12.** Graph showing changes in lava temperature with flow distance for illustrative pahoehoe and aa flows. Inset shows lava flowing through a lava tube.
Data from Cashman *et al.* (1994, 1999).

for typical flow velocities of 1–2 m/s (4–7 km/hr). In contrast, similar flow velocities in open channels result in cooling rates of 15–50°C/hr. As cooling creates crystals within the melt, this comparison shows that crystallization is much more rapid in open channels. Lava tubes form when a continuous surface crust forms across the flow. This is most easily accomplished when flows move slowly, a condition requiring either low effusion rates or flat slopes. As lava tubes are more thermally efficient than open channels, this means that some of the longest lava flows on Earth are those that traverse very flat slopes, an outcome that initially seems counter-intuitive. The same relationship appears to be true on other planets.

Lava emerging from insulated tubes typically forms smooth-surfaced pahoehoe flows. Tubes on moderate slopes are "open" (have an air space above the flow surface; Figures 2.12 and 2.13(a); see color section also), while those on low slopes are "filled". When the tube system fills and pressurizes because of either an increase in lava supply or a blockage downflow, lava may break out to form surface flows. Surface flows advance erratically by rupture of the solidifying surface crust (Figure 2.13(b)). Rapid cooling of the newly exposed hot lava forms a thin surface skin. This skin may wrinkle when compressed against the slow moving flow front to form the ropey surface that is characteristic of pahoehoe lava flows (Figure 2.13(c)). On near-horizontal slopes, pahoehoe flows commonly inflate with time, as lava is added internally from upslope. Slow moving inflated flows from Kilauea volcano consumed the village of Kalapana in 1990, and in so doing provided volcanologists with a detailed view of the inflation process. Morphologic evidence of lava flow inflation has since been recognized in outcrops of "flood basalts," enormous out-pourings of lava that have marked the onset of hot spot activity around the globe.

**Figure 2.13.** Morphologic features of pahoehoe flows. (a) An "open" lava tube with lava flowing well beneath the solidified crust. (b) Small pahoehoe breakout. (c) Actively forming pahoehoe "ropes".
(a) See color section also.

### 2.2.4   Lava flow hazards

Basaltic lava flows rarely pose direct hazards to human life, although in 1977 unusually fluid lava from Nyiragongo volcano, Zaire, advanced at rates of up to 100 km/hr, killing as many as 300 people. More common is substantial damage to property and, in some circumstances, long-term human health problems from protracted emissions of volcanic gases.

Property damage is the most conspicuous consequence of current lava flow activity around the world. For example, a 2001 eruption of Nyiragongo volcano sent fluid lava pouring through the streets of the city of Goma, affecting an estimated 400,000 inhabitants. The eruption that created Parícutin volcano, Mexico, covered 24.8 km$^2$ of land with lava during its nine years of activity, including the towns of Parícutin and San Juan Parangaricutiro. Frequent eruptions of Mount Etna, Sicily, often threaten towns that encircle the volcano. Of particular concern at Mount Etna are eruptions that emerge from the flank of the volcano, in the middle of populated areas, rather than at the summit, which is preserved as a national park. A particularly voluminous flank eruption ($\sim$1 km$^3$) in 1669 engulfed parts of the city of

Catania, including the Castello Ursino, built by Emperor Frederick II in the 13th century.

Mitigation of risks posed by lava flows takes several forms. Frequently active volcanoes such as Kilauea (Hawaii), Mount Etna (Sicily), and Piton de la Fournaise (Réunion Island), have well established monitoring networks. Scientists use these networks to track the upward migration of magma, which is commonly accompanied by swelling (inflation) of the volcano's surface, cracking of near-surface rocks to generate earthquakes, and gas emissions. The goal of such monitoring efforts is to warn public officials and local communities of impending volcanic activity. Once an eruption is underway, both ground-based and remote observations permit tracking of individual flows. When combined with models of expected flow paths and rates of flow advance, these observations can be used to anticipate the arrival of lava flows in specific regions, thus facilitating evacuation and preventing loss of life.

Problems of property loss have been addressed in two ways. Some communities have attempted to control lava flows by either slowing flow advance or diverting flow paths. The first record of such an attempt is from the 1669 eruption of Mount Etna, when lava approached Catania. Local citizens attempted to divert flows away from the city by breaching channel levees and simultaneously blocking the flow within the channel. Although these efforts had only a temporary effect at the time, refinement of these techniques allowed successful diversion of lava flows away from the town of Zafferana Etna during an eruption in 1991–1993. Perhaps the most famous lava diversion effort, however, is the Icelandic fight to save the harbor of Heimaey when it was threatened by an advancing lava flow in 1973. Seawater pumped from the ocean and sprayed on the flow front cooled the lava, preventing flows from covering the narrow harbor entrance. Saving the harbor responsible for 12% of the nation's fisheries income was thus accomplished, although at the expense of part of the town, which was covered by new flows emerging from behind the stalled flow front. Over the longer term, lava flow hazards are best handled by land use planning. To this end, scientists at the USGS have developed lava flow hazard maps for the island of Hawaii based on the frequency of lava flow coverage in a given area (Figure 2.14). Such maps allow city planners to make informed decisions about future development.

Persistent and far-flung hazards from lava flow eruptions result primarily from volcanic gas emissions. The current eruption of Kilauea has created an acid haze of gas, sulfuric acid, and ammonium sulfate aerosols (dispersions of liquid particles in a gas) known as *vog* (volcanic fog). This haze not only lingers over the beautiful beaches of the Kona coast, but also drifts north-west with the prevailing winds to cover all of the Hawaiian islands, limiting visibility and increasing the incidence of respiratory diseases such as asthma. Vog alerts are now issued by the Hawaiian Volcano Observatory in the same way that smog alerts are issued in major cities.

The 1783–1784 eruption of Laki volcano, Iceland, provides a dramatic example of the deleterious effects of volcanic gases. This 10-month eruption produced $15 \, \text{km}^3$ of lava and contributed an estimated 50 million tons of sulfur (as $SO_2$) to the atmosphere, as well as associated acidic gases such as chlorine and fluorine. In the

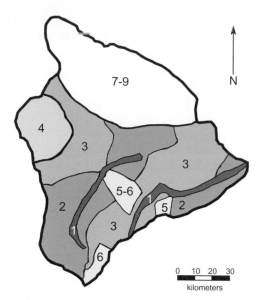

**Figure 2.14.** Lava flow hazard map for the island of Hawaii. Zones are defined as follows: (1) >25% of land area covered since 1800 AD (active rift zones); (2) 15–25% of land area covered since 1800 AD; (3) 1–5% of land area covered since 1800 AD, and 15–75% in the past 750 years; (4) 5% of land covered since 1800 AD, <15% of land area covered in the last 750 years; (5–6) no flows since 1800 AD and variable coverage in the last 750 years—these regions are currently protected from flow inundation by topography; (7–9) no lava flow coverage in the past 750 years.
Modified from USGS Hawaiian Volcano Observatory website; http://pubs.usgs.gov/gip/hazards/maps.html.

words of an Icelandic priest who witnessed the eruption, "... more poison fell from the air than words can describe: ash, volcanic hairs, rain full of sulphur and saltpeter, all of it mixed with sand. The snouts, nostrils and feet of livestock grazing or walking on the grass turned bright yellow and raw. All water went tepid and light blue in color and rocks and gravel slides turned grey. All the Earth's plants burned, withered and turned grey, one after another, as the fire increased and neared the settlements... the foul smell of the air, bitter as seaweed and reeking of rot for days on end, was such that many people, especially those with chest ailments, could no more than half-fill their lungs with air."[6] Ultimately this eruption was responsible for the death of 20% of Iceland's population. Farther away, ash was carried towards Europe, accompanied by a persistent blue haze of acid aerosols and resulting destruction of vegetation. The subsequent three winters were unusually cold in both North America and Europe. The connection between large outpourings of basaltic lava and persistent acid rain have led some scientists to speculate on more

---

[6] Account of the Laki eruption by Rev. Jon Steingrimsson, published in translation in 1998 as *Fires of the Earth: The Laki eruption 1783–1784* by University of Iceland Press and the Nordic Volcanological Institute, Reykjavik, 93 pp.

far-reaching impacts. It has long been recognized that episodes of flood basalt volcanism appear to correlate with times of mass extinction. Although the importance of volcanism relative to other factors (such as meteorite impacts) is still being debated, the eruption of 100s to 1,000s of km$^3$ of basaltic lava would certainly affect the global climate for years if not decades.

## 2.3  SUBDUCTION ZONE VOLCANISM

*Mountain waking*
*Quaking breaking*
*Open to the sky*

*Lightning ripping*
*Through the meadows*
*Ashes blowing high*

*Heavens boiling*
*Hillsides burning*
*Rivers turn to steam*

*Forests hurtling*
*Valleys choking*
*Lakes and spirits scream*[7]

In striking contrast to the normally benign lava flows of Hawaiian volcanism is the more explosive volcanism related to the subduction zones that mark locations around the world where tectonic plates collide (Figure 2.4). Such was the ca. 1600 BC eruption of Santorini volcano on the island of Thera in Mediterranean Greece. This eruption destroyed the island, burying the Minoan city of Akrotiri and causing the local population to flee to neighboring Crete (Figure 2.15). The effects of the eruption—intense earthquakes, ash fall, and tsunamis—were likely felt around the entire Mediterranean. Ancient memories of this disastrous event may be recorded as the "Battle of the Titans" in Hesiod's *Theogony*,[8] as the adventures of Jason and the Argonauts in their return to Greece with the prized Golden Fleece, and even as the plagues of Egypt. Santorini may also have been the model for Plato's Atlantis, an island destroyed by "earthquakes and floods of extraordinary violence, and in a single day and night all your fighting men were swallowed up by the Earth, and the island of Atlantis was similarly swallowed up by the sea and vanished."[9] This description strongly suggests an island destroyed in a catastrophic eruption. Such

[7] From the poem *Remember Spirit Lake*, written by "Cricket" about the 1980 eruption of Mount St. Helens; published in 1981 by the National Speleological Society and reproduced in J.Z. deBoer (2002) *Volcanoes in Human History*. Princeton University Press, Princeton, 295 pp.
[8] D. Wender (1973) *Hesiod and Theognis*. Penguin, London, 170 pp.
[9] Plato (1977) *Timaeus and Critias*. Penguin Books, London, 167 pp.

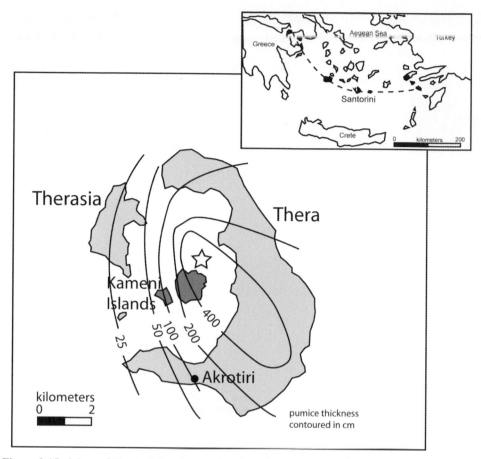

**Figure 2.15.** Map of Santorini volcano showing the remnants of the old edifice (medium gray), vent location for the ca. 1600 BC eruption, contours of pumice fall thickness from that eruption, and the location of post-caldera volcanism (dark gray).
Modified from Druitt *et al.* (1999).

was the 79 AD eruption of Vesuvius volcano near Naples, Italy, responsible for both the demise and the preservation of the cities of Pompeii and Herculaneum (Figure 2.16). While both cities were covered with volcanic pumice, ash, and mud, and thus partially destroyed, the volcanic deposits ultimately preserved the beautiful buildings, mosaics, and frescos of both towns, providing us with a remarkably complete picture of Roman life in the first century AD.

What are subduction zones, and why do they produce such explosive eruptions? In regions of subduction, dense oceanic lithosphere is thrust beneath either another oceanic plate, as in the Aleutians and the Philippines, or continental crust, as in the north-western US and Central and South America (Figure 2.4). The subducting lithosphere contains water-rich minerals—minerals that lose their water as the plate descends into the asthenosphere. Water lost from subducted material rises

**Figure 2.16.** The 79 AD eruption of Vesuvius, Italy. (a) View of Vesuvius from Pompeii. (b) Pumice fall deposit from the 79 AD eruption; wine amphora for scale.

into the overlying lithosphere and causes melting. The melt eventually emerges from volcanoes, cycling heat and water out of the Earth's interior and constructing chains of volcanoes known as "arcs" because of their common arcuate shape (e.g., the Aleutians). Volcanic arcs form parallel to ocean trenches at a location approximately 100–150 km above the subducted slab. Volcanoes are spaced at regular intervals of 30 to 100 km along the arc, probably a consequence of periodic upwelling of magma from the zone of deep melting. Water and other volatile elements ($CO_2$, S, Cl, F)

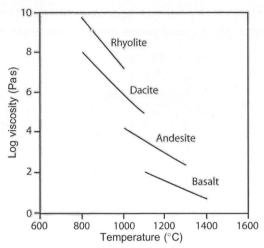

**Figure 2.17.** Variation of viscosity with temperature for common magma compositions.
Redrafted from Spera (2000).

contributed by the subducted slab are responsible for the explosivity of arc volcanoes. Volcanic rocks formed from these melts are typically higher in $SiO_2$ than basaltic magma erupted at hot spots, and are termed andesite ($SiO_2 = 57$–$63\%$), dacite ($SiO_2 = 63$–$70\%$), or rhyolite ($SiO_2 > 70\%$). These cooler, more silica-rich magmas from subduction zone volcanoes are much more viscous (less fluid) than basaltic lavas (Figure 2.17).

Eruptions of subduction zone volcanoes are typically less frequent than those of basaltic volcanoes, but tend to be both larger and more explosive. This inverse relationship between eruption frequency and magnitude can be quantified using the *Volcano Explosivity Index* (VEI), a logarithmic scale that classifies eruptions using a combination of erupted volume and intensity (mass eruption rate). A plot of VEI against frequency of eruptions (also plotted on a logarithmic scale) yields a straight line (Figure 2.18). From this we can see that volcanic eruptions the size of the Vesuvius eruption of 79 AD (VEI = 5) occur about once per decade, while eruptions the size of Santorini (VEI = 6) occur only about once per century. Note that very large eruptions (VEI = 8) occur in the geologic record only about once every 100,000 years, much less frequently than predicted by simple extrapolation of the straight line. Do these eruptions truly not fit the pattern, or is the geologic record of these events too scant? We do not know the answer to this question.

### 2.3.1  Volcano morphology

In contrast to the low slopes and comparatively smooth surfaces of Hawaiian shield volcanoes, slopes of composite volcanoes in arc environments are steep and constructed of intermingled lava flows and fragmental deposits (Figure 2.19(a)). Lava flows on these volcanoes are thick (10–50 m), with rough blocky surfaces. Fragmen-

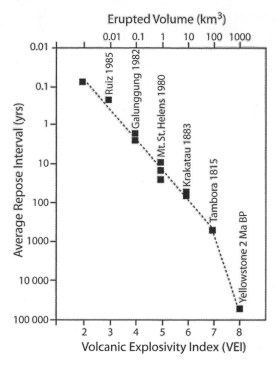

**Figure 2.18.** Average time between volcanic eruptions of different magnitudes, as defined by the VEI (defined primarily by eruptive volume).
Redrafted from Simkin and Siebert (2000).

tal deposits are composed of highly vesicular pumice, fine-grained volcanic ash, blocks of dense lava, and mixtures of old and new rock transported by volcanic mudflows. The symmetrical shape of many composite volcanoes results from eruptions out of central summit vents. This symmetry may be marred by eruptions from satellite vents or by the scars of destructive events, such as the catastrophic landslide and lateral explosion that initiated the 1980 eruption of Mount St. Helens, Washington, U.S.A. (Figure 2.19(b)). Composite cones themselves are long-lived (~500,000 years) but rapid cone growth may occupy only a short part of that protracted history (<100,000 years). A complex history is thus written in each edifice, a story with an infinite number of variations but with common and repeated themes.

The shapes of composite volcanoes may be viewed as a balance between processes that build cones and those that destroy them. Constructive processes include eruptions of lava flows and domes, as well as explosive events that deposit thick sequences of pumice and ash. These processes create cone-shaped volcanoes, with heights that are proportional to the edifice volume (Figure 2.19(c)). Composite volcanoes rarely exceed 3,000 m in height from their base, and cone volumes are similarly restricted to <~200 km$^3$, far smaller than the massive volumes of basaltic

**Figure 2.19.** Morphological features of composite cones. (a) Augustine volcano, Alaska, summit done flanked by fans of fragmental debris (pyroclastic flows and debris flows); Augustine last erupted in 1986. (b) View of Mount St. Helens from the north-west, showing the amphitheater created by a catastrophic failure of the volcano's north flank on 18 May, 1980. (c) Symmetrical cone of Tungurahua volcano, Ecuador, emerging from a low cloud layer. (d) Subsidiary cone Crater Peak on Alaska's Mount Spurr volcano. Crater Peak last erupted in 1992. (e) Pumice and ash deposits from Pululahua volcano, Ecuador. (f) Active Noboribetsu geothermal field in Hokkaido, Japan, illustrating typical hydrothermal alteration.

shields. Similar volcano heights and volumes of composite cones around the world suggest that there is a physical limit to volcano size. This limit is probably related to the maximum pressures that can develop in magma reservoirs at depth, pressures required to trigger eruptions. When a volcano's summit is too high, it is simply not possible to build up sufficient pressure at depth to force the magma to the surface. When this height limit is reached, magma may emerge through satellite vents on the volcano's flanks (Figure 2.19(d)) or the edifice may fail. Volcanic edifices are destroyed continuously by erosion and catastrophically by flank failure. Erosion rates are controlled over long time periods by climate, but over short periods by the recent eruptive history. Explosive eruptions scatter large amounts of easily eroded pumice and ash over the landscape (Figure 2.19(e)). This pyroclastic material is rapidly eroded from steeper slopes and deposited in lowland valleys, particularly during seasonal rains. Alteration of volcanic rocks by acid hydrothermal waters (Figure 2.19(f)) may also weaken central edifices, making them susceptible to collapse during minor eruptions, heavy rains, or tectonic earthquakes. Alternating cycles of growth and destruction means that only an incomplete eruptive history is preserved in any single composite volcano.

A discussion of the morphology of composite volcanoes is incomplete without mention of calderas, large near-circular depressions formed by the collapse of a volcano's summit as the result of wholesale evacuation of the underlying magma reservoir. A classic example of a caldera is Crater Lake, Oregon (Figure 2.20(a)), formed in a large (VEI = 6–7) eruption almost 8,000 years ago. Calderas may range in size from 2 to >50 km in diameter, with the size generally proportional to the volume of magma erupted (from $\sim$5 km$^3$ for the smallest calderas to $\sim$2,000 km$^3$ for the largest). Smaller calderas usually form within large volcanic edifices, as seen at Crater Lake. In contrast, large calderas form "inverse volcanoes", large depressions that extend beyond any edifice structure and are often occupied by voluminous lakes such as Lake Taupo, New Zealand (Figure 2.20(b)) and Lake Toba, Sumatra. The collapse origin of calderas was recognized in the mid-nineteenth century by the careful field observations of Lyell at La Palma, Canaries; Dutton at Kilauea caldera, Hawaii; Fouque at Santorini volcano in the Aegean; and importantly, by Verbeek's study of the 1883 eruption of Krakatau volcano in Indonesia.[10] These studies showed that the amount of lithic material (dense rocks from the old summit of the volcano) ejected during caldera-forming eruptions was much less than the volume of the depression left behind. Caldera formation thus requires ejection of magma from below, and collapse of the edifice into the resulting space. Collapse may occur in different ways, from piston-like downward motion of the entire crater floor to piecemeal collapse of different floor pieces at different times, from trap-door asymmetrical collapse to simple downward sagging and to funnel-shaped depressions that merge towards excavated craters typical of small volume eruptions (Figure 2.20(c)).

[10] R.D.M. Verbeek (1885) *Krakatau*. Batvia, 495 pp. Reprinted in T. Smith and R.S. Fiske (1983) *Krakatau 1883: The Volcanic Eruption and Its Effects*. Smithsonian Institution Press, Washington, DC, 464 pp.

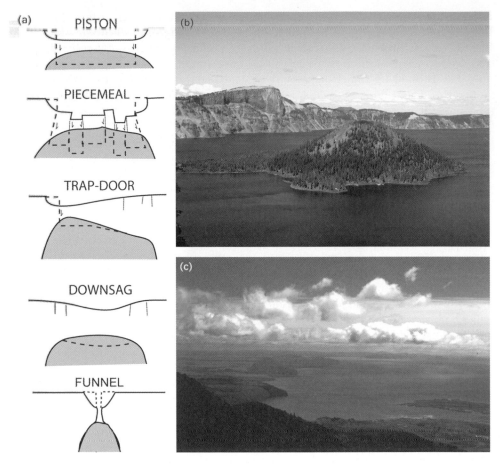

**Figure 2.20.** Morphology of calderas. (a) Processes of caldera collapse. (b) Crater Lake caldera, Oregon, USA. (c) Lake Taupo (Taupo caldera), North Island, New Zealand.
(a) Redrafted from Lipman (2000).

### 2.3.2   Explosive eruptions

Composite volcanoes are best known for producing towering columns of gases and particles during *Plinian* eruptions. This eruption style is named after the Roman Pliny the Younger, who provided us with the first description of this phenomenon in his account of the 79 AD eruption of Vesuvius volcano, Italy. He described the eruption column as, "a pine-tree, for it shot up a great height in the form of a trunk, which extended itself at the top into several branches...",[11] an apt description of Italian umbrella pines. Pliny's account goes on to include several key components of a typical Plinian eruption. First there was a succession of earthquakes,

[11] **M. Krafft (1993)** *Volcanoes: Fire from the Earth.* Harry Abrams, Inc, New York, 207 pp.

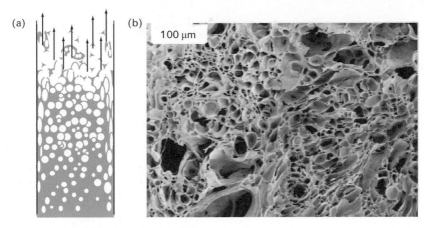

(a)                (b)   100 µm

**Figure 2.21.** (a) Magmatic fragmentation occurs by rapid bubble nucleation and growth (vesiculation) followed by disruption of bubbly material. (b) Electron microscope image of pumice from Crater Lake, Oregon, showing that pumice is formed of numerous very small bubbles connected by thin films of glass (formerly melt).

"so violent that one might think that the world was not being merely shaken, but turned topsy-turvy." Then the beginning of the eruption and generation of a towering cloud of pumice and ash, "a black and dreadful cloud bursting out in gusts of igneous serpentine vapour (that) now and again yawned open to reveal long fantastic flames, resembling flashes of lightning but much larger." As this material rained back to Earth, "there was a deeper darkness that prevailed than in the most obscure night." With time, the vent feeding the high column widened, creating hot flows of pumice and ash that appeared as a "dark cloud" that "began to descend upon the Earth, and cover the sea." It was these flows that inundated the towns of Pompeii and Herculaneum, causing massive destruction. As the flows encountered the nearby ocean, "the sea sucked back, as if it were repulsed by the convulsive motion of the Earth", creating the local tsunamis common to volcanoes whose flows are within reach of water.

What physical mechanisms are responsible for these events? Magma generated at subduction zones rises through the Earth's crust and accumulates at depths of 5–10 km below a volcano's summit, eventually generating sufficient pressure to force its way toward the surface. Rock breakage during this upward passage causes earthquakes that precede large eruptions by days, weeks, or months. As the magma rises, dissolved gases exsolve from the melt to form bubbles. As the bubbles grow, the bubble–melt mixture expands and accelerates toward the surface. When bubbles within the magma occupy a sufficiently large volume (approximately 75%) or when the rise rate is sufficiently high, the magma fragments, creating discrete clots of bubbly melt within a hot and rapidly expanding gas (Figure 2.21(a)). This mixture emerges from the vent at close to the speed of sound and continues to expand upward, engulfing the surrounding air and heating it to fuel the growing plume.

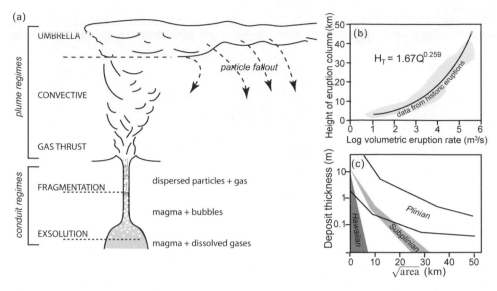

**Figure 2.22.** (a) Cartoon showing important regions in volcanic conduits and plumes. The *exsolution* depth marks the first formation of bubbles in the melt. The *fragmentation* depth is the point at which the melt transforms to an accelerating gas phase. The *gas thrust* region delimits the part of the plume driven by the momentum of the magma exiting the vent. In the *convective* region of the plume, the plume rises buoyantly, reaching the height of neutral buoyancy as it enters the *umbrella region*, where it extends in the prevailing wind direction and deposits particulate matter. (b) Graph showing the relationship between eruption column height and volumetric eruption rate. (c) Graph illustrating the variation in the pattern of fallout (in the form of thickness vs. distance, as $\sqrt{\text{area}}$) for Hawaiian, sub-Plinian, and Plinian eruptions.
(a) Modified from Cioni (2000); (b) modified from Sparks *et al.* (1997); (c) modified from Houghton *et al.* (2000).

Bubbly magma clots cool during transport to form low-density pumice (a rock that floats in water; Figure 2.21(b)).

   After exiting from the vent, a Plinian eruption column rises until its density matches that of the surrounding atmosphere (it becomes neutrally buoyant), at which point the plume spreads and elongates in the direction of the prevailing wind (Figure 2.22(a)). The height of the eruption column is proportional to the volumetric eruption rate (Figure 2.22(b)). Column height, in turn, controls the distribution of pumice and ash that fall from the plume (higher columns transport volcanic material further). As individual particles of pumice and ash fall from the plume, each particle follows a trajectory determined by its position in the column and its fall velocity. Resulting pumice deposits mantle the landscape, covering large areas in pumice and ash (Figure 2.19(e)). Both the thickness (Figure 2.22(c)) and the average particle size of these deposits decrease away from the volcano. Characterization of pumice fall deposits (i.e., measurement of deposit thickness particle sizes) can be used to infer the volume of erupted material, as well as the rate of eruption (by relating the fall deposit to column height), and the prevailing

**Figure 2.23.** Caldera collapse at Crater Lake, Oregon. (a,b) Cartoons illustrating the Plinian column and caldera collapse phases of the eruption. (c) Large boulder of lag breccia (part of the volcanic edifice) marking the time of caldera collapse. It lies at the boundary between the pumice fall deposit produced during the high column phase of the eruption and the pyroclastic flow deposit produced during caldera collapse. (d) Pyroclastic flow deposits from Crater Lake at The Pinnacles, Crater Lake National Park, Oregon.
(a,b) Modified from the USGS web site: http://pubs.usgs.gov/gip/hazards/maps.html.

wind direction (by looking at the map view shape of the deposit relative to its origin—low density particles will travel away from the volcano in the direction of wind transport; Figure 2.15). In this way, deposits from ancient eruptions may be used to infer past eruptive activity at individual volcanoes, information that is often a useful guide to possible future activity.

Vigorous eruption columns change their character with time as the energy of the explosive mixture erodes the vent region and magma withdrawal destabilizes the edifice. This occurs as eruption through a single central vent (Figure 2.23(a)) causes cracking around a larger volume of the edifice. Eventually, as support is withdrawn from below, the region around the vent collapses (Figure 2.23(b)), hurling out large blocks of the old edifice together with large volumes of pumice and ash (Figure 2.23(c)). Collapse has the immediate effect of widening the eruptive vent, the consequences of which are twofold. First, vent widening causes an increase in the eruption rate and a partial or complete transition in eruptive behavior from a high buoyant column to dense, hot, fast, and highly destructive flows of particles and gases (Pliny's dark cloud). These pyroclastic flows are often funneled down valleys. Initially, they accelerate by engulfing and heating the air that they encounter, ceasing only when the magmatic heat is no longer sufficient to drive the flow. Pumice and ash deposits of such flows are called ignimbrites, a term derived from the Greek meaning "fire cloud rock" (Figure 2.23(d)). Fine ash winnowed from the flowing mixture may rise along with the hot air emerging from the flow, forming secondary plumes that may approach the height of the original Plinian columns. When additional ash is created by interaction of hot pyroclastic flows with ocean water, secondary "co-ignimbrite" ash may comprise a large part of an eruptive deposit. For example, pyroclastic flows created during the large (VEI = 7) eruption of Tambora volcano, Indonesia, in 1815, produced co-ignimbrite ash that constituted an estimated 80% of the total erupted volume of $\sim$75 km$^3$. During this eruption, large volumes of ash and aerosols were injected into the stratosphere, dramatically affecting the global climate. The year of 1816 was known as the "year without a summer" in New England. Summer in Europe was also unusually cold and wet, with temperatures as much as 10°C below normal. Not coincidentally, it was in the summer of 1816 that the poet Lord Byron, who was vacationing in northern Italy, wrote his poem *Darkness*:[12]

> *I had a dream, which was not all a dream.*
> *The bright sun was extinguished, and the stars*
> *Did wander darkling in the eternal space,*
> *Rayless, and pathless, and the icy Earth*
> *Swung blind and blackening in the moonless air;*
> *Morn came and went—and brought no day,*
> *And men forgot their passions and their dread*
> *Of this their desolation; and all hearts*
> *Were chilled into a selfish prayer for light . . .*

Ironically, as Lord Byron did not know the origin of this temporary darkness and cold, he continued with

> *Happy were those who dwelt within the eye*
> *Of the volcanoes, and their mountain torch . . .*

More famous than Byron's poem, however, is another piece of literature that can be attributed to the Tambora eruption. In response to Lord Byron's suggestion that he and his friends the Shelleys write ghost stories to amuse themselves while trapped indoors because of the gloomy summer weather, Mary Shelley created the most famous ghost story of all time, *Frankenstein*.

[12] G. Gordon, Lord Byron (1972) *Selected Works*. Holt, Rinehart and Winston, Inc., New York, 698 pp.

### 2.3.3   Silicic lava flows and domes

While Plinian eruptions represent the most dramatic eruptive style of composite volcanoes, more common are the smaller and less violent explosions that accompany the eruption of lava flows and domes. Explosions during such eruptions may be sustained or may occur in short pulses, ranging from *sub-Plinian* (VEI = 3–4) to brief violent explosions of $\ll 0.1\,km^3$ known as *Vulcanian*. These explosions commonly disrupt shallow lava plugs or domes, creating dense pyroclastic flows that are termed *nuées ardentes*, or "glowing clouds", based on Lacroix's 1904 descriptions of such flows from Mount Pelee, Martinique.[13] Less common are explosions related to cryptodomes, shallow magma intrusions such as that which triggered the 1980 eruption of Mount St. Helens. Here intrusion of magma into the volcanic edifice created an enormous bulge on the northern flank of the volcano. On 18 May, 1980, the bulge failed in a large landslide $(0.67\,km^3$; Figure 2.24(a)) that, in turn, depressurized the cryptodome to generate a powerful lateral blast, or pyroclastic surge. A similar pyroclastic surge from the volcano Mount Pelee (1902) killed more than 25,000 inhabitants of the town of St. Pierre, at a distance of 7 km from the volcano (Figure 2.24(b)). More recently, violent explosions have punctuated the intermittent growth of lava domes at Unzen volcano, Japan, and at the Soufriere Hills volcano, Montserrat. Deposits from these eruptions are of two kinds. Energetic surges carry mostly fine ash, and leave only thin deposits in the affected area. More limited in extent are thick block-and-ash flows, mixtures of large boulders and finely pulverized rock deposited in channels traveled by individual pyroclastic flows.

Lava flows from arc volcanoes vary over a wide range in both composition and morphology. Andesitic lava flows often have pronounced lateral levees that may be several meters high (Figure 2.25(a,b)). Like Hawaiian lava flows, they are typically long relative to their width, except on very flat slopes. In contrast, rhyolitic obsidian flows are thick (10s of meters) with stubby flows and no channel development, although compressional folds are well preserved on flow surfaces (Figure 2.25(c)). Lava domes form when lava viscosity is high and eruption rates are low. Most domes are therefore composed of lava rich in $SiO_2$, such as andesite, dacite, and rhyolite, because of their relatively high viscosities. Slow ascent allows the magma to lose some of its gases non-explosively. As water is lost, the magma starts to crystallize, creating pasty lava that creeps out of summit vents and flows only short distances (Figure 2.26(a,b)). Lava domes range in size from tens of meters across to kilometers, and may reach 100s of meters in height. Some domes have rough scoriaceous surfaces, while others are composed of large tumbled blocks or pierced by large vertical blocks of solid lava (Figure 2.26(c,d)). Such vertical blocks are called *Pelean spines*, in reference to the eponymous growth of a spectacular spine that grew to 300 m in height in the months following the 1902 eruption of Mount Pelee. As in basaltic lava flows, most of these morphological features—round or elongate, rough or smooth—can be explained by variations in the relative rates of flow advance (spreading) and of cooling and crystallization.

---

[13] A. Lacroix (1904) *La Montagne Pelee et ses eruptions*. Masson, Paris.

**Figure 2.24.** Products and consequences of explosive eruptions from growing lava domes and cryptodomes. (a) View out of the crater of Mount St. Helens showing pyroclastic flow deposits from eruptions that occurred in the summer of 1980 (foreground) and hummocky topography of the landslide that initiated the catastrophic eruption on 18 May, 1980. (b) Ruins of an estate outside the town of St. Pierre, Martinique, destroyed on 8 May, 1902, by an eruption of Mount Pelee.

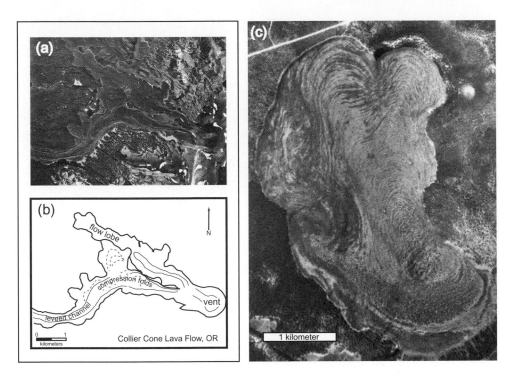

**Figure 2.25.** Lava flow morphologies typical of arc-type volcanoes. (a) Aerial photo showing part of a young (ca. 1500 years) andesitic lava flow in the central Oregon Cascades. (b) Interpretive sketch of the aerial photo. The vent is a prominent cinder cone at the far right of the image. Note that flow lobes are much longer than they are wide, that flow levees are well developed (seen as straight lines in the flow at the bottom left of the image), and that surface folds are well developed behind stalled flow fronts. (c) Aerial photo of the Big Obsidian flow, Newberry volcano, Oregon. The vent is the mound-like feature toward the bottom of the image. Note the poorly developed flow lobes and well developed surface folds.

### 2.3.4　Hazards associated with subduction zone volcanism

Hazards associated with subduction zone volcanism are numerous and often deadly. Volcanic avalanches, hot pyroclastic flows, and volcanic mudflows destroy everything in their path, while flows that enter the ocean may generate tsunamis that flood nearby coastal regions. Heavy ash fall makes breathing difficult, and accumulation of ash may collapse roofs. Volcanic gases may kill directly (especially when rich in $CO_2$) or indirectly, by short-term acid contamination of fields and crops or long-term changes in the climate. And in our increasingly technological world, volcanic eruptions pose new hazards—ash interference with aircraft, disturbance of reservoir and hydroelectric power schemes by volcanic mudflows, and economic disruption when volcanoes that are tourist destinations turn violent. Over the past 300 years, 260,000 people have fallen victim to volcanic eruptions. While our ability

**Figure 2.26.** Morphology of lava domes. (a) Summit of Augustine volcano, which is composed of several recent lava domes. (b) Summit dome of Merapi volcano, Indonesia. Again note the very steep slopes and the individual thick flows and domes that comprise the summit. (c) Aerial shot of the summit of Augustine volcano, Alaska, showing parts of three separate domes (from right to left, remnants of domes from 1963, 1976, and 1986). The last lava erupted in 1986, producing a small lava spine seen protruding from the youngest dome. (d) Spine formed in 1983 from the growing dome of Mount St. Helens.

to monitor and predict the course of volcanic activity has improved dramatically over the past decades, a major challenge for the future is mitigation of the risk of volcanic eruptions, particularly in the face of the rapid increase in populations that inhabit volcanic regions.

Perhaps the most dramatic of volcanic hazards are those presented by pyroclastic flows and surges—energetic hot flows of gas and particles that strip landscapes of vegetation and structures in a matter of minutes. Pyroclastic flows may travel at speeds of up to 150 m/s (300 mph), and may have temperatures up to 600–700°C. People die quickly when overcome by hot ash clouds, either by asphyxiation and/or burning. Little can be done to mitigate the hazards posed by these events except evacuation, which requires accurate and timely prediction of volcanic eruptions. As devastating pyroclastic flows often occur at the culmination of eruptive activity, it is often possible for scientists to provide some warning of these events. When pyroclastic flows travel down snow-clad volcanic slopes, or when they mix with water in river drainages, they produce volcanic flows of mud and rock known by the Indonesian term *lahar*. Flow velocities of 13–30 km/hr are common, and lahars may reach speeds of up to 180 km/hr. The high densities and velocities of these flows make them very erosive, such that they "bulk up" with the addition of sediment during flow downstream. Most notorious in the 20th century were debris flows produced during a relatively small (VEI = 3) eruption of Nevado del Ruiz, Columbia, in 1985, that covered the town of Armero and more than 23,000 of its inhabitants. This event was particularly unfortunate as the hazard was known, and loss of life could have been largely avoided by implementation of effective warning and evacuation strategies.

Volcanic tsunamis have also been responsible for many volcano-related deaths where either large avalanches (e.g., Unzen 1792) or pyroclastic flows (Santorini 1650 BCE; Krakatau 1883) have entered the ocean. Volcano-generated tsunamis may affect areas that do not receive heavy ash fall or pyroclastic flows, and thus may arrive without warning. A vivid description of a close encounter with a volcanic tsunami is that of Controller P.L.C. Le Sueur, who was living in Beneawang, Indonesia, at the time of the 1883 eruption of Krakatau: "... Suddenly the water returned... I said a quick prayer, asking for help for myself and everybody and prepared myself for death. The water took hold of me, turned me around and threw me away with a terrible force. Then I got stuck between two floating houses. I couldn't breathe any more and I thought the end had come. But suddenly they parted and I got hold of a banana trunk and stuck to it with all my strength. I don't know how long I floated around, but again the water returned to the sea and once again I stood on solid ground. Again I sat there for at least an hour without moving and it was dark everywhere and the mud rain was still going on."[14] Similar tsunamis affected much of the Aegean Sea after the Santorini eruption, possibly reaching heights of up to 9 m on Crete, with waves only slightly smaller on mainland Greece and Turkey.

While not necessarily life-threatening, heavy ash fall from eruption columns creates an impenetrable darkness and suffocating environment that is terrifying for all who experience it. Memories of "times of darkness" are often the most persistent remnants of spoken traditions linked to eruptions which occurred long ago. When ash fall is sufficiently heavy, accumulations of ash may collapse roofs. Problems of ash accumulation are exacerbated by the addition of atmospheric water to the falling ash, creating muddy rain that is heavier than ash alone. A 15 cm (6 inch) accumulation of wet ash generates a load of $\sim 200 \, kg/m^2$, greatly in excess of normal building codes for roof loads. As ash blankets the landscape, it can also cause immediate damage to crops and livestock, although in the long term volcanic ash adds minerals to the soil that may enhance its productivity. A final hazard of ash relates to its deleterious effect on airplanes—in the past two decades more than 100 jet airplanes have flown into volcanic ash clouds. Ash is abrasive and can cause complete frosting of window exteriors. Ingestion of volcanic ash into aircraft engines can cause melting

[14] T. Simkin and R.S. Fiske (1983) *Krakatau 1883*. Smithsonian Institution Press, Washington, DC, 464 pp.

and resolidification, and in the most severe cases may lead to engine failure. Airplanes may encounter ash far from the source volcano. As ash clouds cannot be detected by on-board instruments, Volcanic Ash Advisory Centers (VAACs) have been established to monitor and communicate regional levels of volcanic activity to meterological and civil aviation organizations.


## 2.4  SUMMARY

*The creation of myth must continue as long as Kilauea continues to erupt.*[15]

Volcanic activity is a fundamental Earth process. A result of thermal convection driven by heat from the Earth's core, it is the primary criterion by which our planet is deemed "alive" rather than "dead", in a geologic sense. This sense has been extrapolated to other planets, hence the excitement when active volcanism was first recognized on Jupiter's moon Io. Our ability to monitor and study processes at active terrestrial volcanoes provides us with an invaluable tool for interpreting not only the past volcanic history of our own planet, but also the processes responsible for surfacing other planets. Specifically, known relationships among lava flow morphology, composition, and conditions of eruption on Earth provide a means of inferring both the composition and eruption rate of planetary flows. Conversely, the excellent preservation of flow surfaces on planets that lack our hydrosphere and atmosphere may provide new insights into emplacement conditions of enormous basalt flows that have flooded Earth's landscape in the past.

Volcanoes have also played a crucial role in human history, although the full extent of that role has yet to be determined. Early in the Earth's history, volcanic activity transferred volatile elements from the Earth's interior to the oceans and atmosphere. Volcanism helped to construct continents, and it is through volcanoes that water and other volatile elements are continuously recycled from the surface to the mantle and back to the surface. Large floods of basaltic lava have punctuated the past several hundred million years of Earth's history, and may have played a role in major life extinction events. 74,000 years ago, a massive (VEI = 8) eruption of Toba, Sumatra, may have sufficiently disturbed the global climate to cause a "bottleneck" in human evolution; it has been suggested that at this time the human population may have been reduced to less than 10,000 individuals. Although such connections are speculative, it is certain that catastrophic eruptions have not only had an enormous impact on local cultures, but have also affected far distant regions, at least for periods of years. Knowledge of past events has traditionally been preserved locally as spoken traditions and mythologies about the landscape and its history. As populations become increasingly concentrated in urban environments, local memories are lost. Our challenge for the future is to cultivate and extend local knowledge of the devastating power of past eruptions while striving to improve our fundamental understanding of volcanic activity, with the ultimate goal of improving

[15] P.K. Kanahele (2001) *Holo Mai Pele*. Native Books, Inc., Honolulu, 67 pp.

the resilience of growing global populations to the effects of ongoing volcanic activity.

## 2.5 REFERENCES

Cashman, K.V., Thornber, C.R., and Kauahikaua, J.P. (1999) Cooling and crystallization of lava in open channels, and the transition of pahoehoe lava to aa. *Bull. Volcanol.*, **61**, 306–323.

Cashman, K.V., Mangan, M.T., and Newman, S. (1994) Surface degassing and modifications to vesicle size distributions in Kilauea basalt. *J. Volcanol. Geotherm. Res.*, **61**, 45–68.

Cioni, R., Marianelli, P., Santacroce, R., and Sbrana, A. (2000) Plinian and subplinian eruptions. In: H. Sigurdsson (ed.), *Encyclopedia of Volcanoes*. Academic Press, San Diego, pp. 477–494.

Druitt, T.H., Edwards, L., Mellors, R.M., Pyle, D.M., Sparks, R.S.J., Lanphere, M., Davies, M., and Barriero, B. (1999) *Santorini Volcano*. The Geological Society, London, 165 pp.

Holcomb, R.T. (1987) *Eruptive History and Long-term Behavior of Kilauea Volcano* (Professional Paper 1350). US Geological Survey, Reston, VA, pp. 261–350.

Houghton, B.F., Wilson, C.J.N., and Pyle, D.M. (2000) Pyroclastic fall deposits. In: H. Sigurdsson (ed.), *Encyclopedia of Volcanoes*. Academic Press, San Diego, pp. 555–570.

Kauahikaua, J.P., Sherrod, D., Cashman, K., Heliker, C., Hon, K., Mattox, T., and Johnson, J. (2003) *Hawaiian Lava-flow Dynamics During the Puu O-o-Kupaianaha Eruption: A Tale of Two Decades* (Professional Paper 1676). US Geological Survey, Reston, VA, pp. 63–87.

Kilburn, C.R.J. (2000) Lava flows and flow fields. In: H. Sigurdsson (ed.), *Encyclopedia of Volcanoes*. Academic Press, San Diego, pp. 291–306.

Simkin, T. and Siebert, L. (2000) Earth's volcanoes and eruptions: An overview. In: H. Sigurdsson (ed.), *Encyclopedia of Volcanoes*. Academic Press, San Diego, pp. 249–262.

Sparks, R.S.J., Bursik, M.I., Carey, S.N., Gilbert, J.S., Glaze, L.S., Sigurdsson, H., and Woods, A.W. (1997) *Volcanic Plumes*. Wiley, Chichester, UK, 574 pp.

Spera, F.J. (2000) Physical properties of magmas. In: H. Sigurdsson (ed.), *Encyclopedia of Volcanoes*. Academic Press, San Diego, pp. 171–190.

## 2.6 BIBLIOGRAPHY AND FURTHER READING

### General

Sigurdsson, H. (2000) *Encyclopedia of Volcanoes*. Academic Press, San Diego, 1415 pp.

### Basaltic volcanism

Chester, D.K., Duncan, A.M., Guest, J.E., and Kilburn, C.R.J. (1985) *Mount Etna: The Anatomy of a Volcano*. Stanford University Press, Stanford, 404 pp.

Decker, R.W., Wright, T.L., and Stauffer, P.H. (1987) *Volcanism in Hawaii* (Professional Paper 1350). US Geological Survey, Reston, VA.

Heliker, C., Swanson, D.A., and Takahashi, T.J. (2003) *The Puu O-o-Kupaianaha Eruption of Kilauea Volcano, Hawaii: The First 20 Years* (Professional Paper 1676). US Geological Survey, Reston, VA, 206 pp.

Macdonald, G.A., Abbott, A.T., and Peterson, F.I. (1983) *Volcanoes In The Sea. The Geology of Hawaii*. University of Hawaii Press, Honolulu, 517 pp.

Rhodes, J.M., and Lockwood, J.P. (1995) *Mauna Loa Revealed; Structure, Composition, History, and Hazards* (Monograph 92). American Geophysical Union, Washington, DC, 348 pp.


## Explosive volcanism

Druitt, T.H. and Kokelaar, B.P. (2002) *The Eruption of Soufriere Hills Volcano, Montserrat From 1995 to 1999* (Geological Society Memoirs No. 21). The Geological Society, London, 645 pp.

Lipman, P.W. and Mullineaux, D.R. (1981) *The 1980 Eruptions of Mount St. Helens, Washington* (Professional Paper 1250). US Geological Survey, Reston, VA, 843 pp.

Newhall, C.J. and Punongbayan, R.S. (1996) *Fire and Mud: Eruptions and Lahars of Mount Pinatubo, Philippines*. Philippine Institute of Volcanology and Seismology, Quezon City, and University of Washington Press, Seattle, 1126 pp.


## Volcanoes and humans

Blong, R.J. (1982) *The Time of Darkness: Local Legends and Volcanic Reality in Papua New Guinea*. University of Washington Press, Seattle, 257 pp.

deBoer, J.Z. and Sanders, D.T. (2002) *Volcanoes in Human History: The Far-Reaching Effects of Major Eruptions*. Princeton University Press, Princeton, 295 pp.

Greene, M.T. (1992) *Natural Knowledge in Preclassical Antiquity*. Johns Hopkins University Press, Baltimore, 182 pp.

McGuire, W.J., Griffiths, D.R., Hancock, P.L., and Stewart, I.S. (2000) *The Archaeology of Geological Catastrophes* (Geological Society Special Publication No. 171). The Geological Society, London, 417 pp.

McPhee, J.A. (1989) *The Control of Nature*. Farrar, Straus & Giroux, New York, 272 pp.

Scarth, A. (1999) *Vulcan's Fury: Man Against the Volcano*. Yale University Press, New Haven, 299 pp.

Simkin, T. and Fiske, R.S. (1983) *Krakatau 1883: The Volcanic Eruption and its Effects*. Smithsonian Institution Press, Washington, 464 pp.

# 3

# Submarine volcanoes: The hidden face of the Earth

*Tracy Gregg* (The State University of New York at Buffalo)

*From the time I was about 4 years old, when people asked me what I wanted to be when I grew up, I would respond with "A doctor or an astronaut". As I grew older, I realized that I didn't know any astronauts, but I did know several doctors. Throughout my public school education, I insisted that I would go to college, major in biology or biomedicine, and then go to medical school; my only doubt was just what type of doctor I would be. When I began my freshman year in college, I dutifully registered for maths, biology, chemistry, and, for fun, a geology course entitled "Earth, Moon, and Mars" taught by James W. Head III at Brown University. That course changed my life, and I whole-heartedly threw myself into the study of geology. I knew I had to go to graduate school after college, if only because I felt that there was so much I still needed to learn before I could call myself a geologist! I went to Arizona State University to study volcanism on other planets. I rapidly became frustrated: the only other planet in the solar system we have collected rocks from is the Moon, and I was trying to understand volcanoes on Mars and Venus. Quite by accident, I stumbled upon the study of volcanism at Earth's mid-ocean ridges. This was the perfect step for me: I could investigate volcanoes in strange, remote environments, but could still collect rocks and bring them back for study! I am now a professor at the University at Buffalo where I continue to investigate volcanoes wherever I can find them in our solar system.*

## 3.1 INTRODUCTION

Over 75% of the Earth's surface is covered with water, and is difficult, expensive, and sometimes dangerous to investigate. Thus, most of what we know about volcanism on Earth is based on studies of *subaerial* (land-based) volcanoes. There are probably around 500–600 active volcanoes on land on Earth today—but that number is undoubtedly dwarfed by the number of active volcanoes quietly erupting beneath the ocean. We don't know how many active volcanoes there are in the oceans, nor do

**Figure 3.1.** The Atlantic Ocean as it would appear if drained of water.
Courtesy of NOAA.

we know precisely how they behave, because no one has ever observed an active eruption on the deep sea floor. We do know for certain that the vast majority of the Earth's volcanic activity occurs under water.


## 3.2   MID-OCEAN RIDGES

Laying the trans-Atlantic telegraph cable in the middle of the 1800s revealed a shallow region—"Telegraph Plateau"—in the center of the Atlantic Ocean. Initial work by Maurice Ewing and Bruce Heezen (both of the Lamont–Doherty Geological Observatory) in the post-World-War-II era (in a US government-funded study to seek out the best potential hiding place for submarines) finally revealed the true character of the Earth's mid-ocean ridges (Heezen *et al.*, 1959; Ewing and Heezen, 1960). Their efforts revealed a huge mountain chain—up to 4 km tall and 10–30 km wide—running down the middle of the Atlantic Ocean (Figure 3.1). When geologists make an observation but don't want to cloud it with their (potentially incorrect) interpretations, they call it like they see it. This mountain range was called the Mid-Atlantic Ridge.

Later, in the 1960s, data compiled from around the globe began to demonstrate

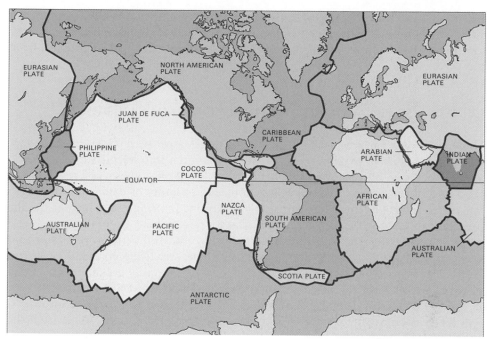

**Figure 3.2.** The largest tectonic plates on Earth. There may be as many as 20 or so mid-ocean ridge plates more than are depicted here, depending on how one defines a "plate".
Image courtesy of the USGS.

the existence of what we now call plate tectonics on Earth (Dietz, 1961; Hess, 1962). In the paradigm of plate tectonics, the surface of the Earth is not one solid shell, but is broken up into a series of puzzle pieces, called plates (Figure 3.2). Plates interact with each other in one of the following ways: (1) they may move away from each other, known as a "divergent boundary"; (2) they may slide past each other, creating a "transform boundary" such as the San Andreas Fault in southern California; or (3) they may crash into each other, creating a "convergent boundary". It turns out that the Mid-Atlantic Ridge (and other similar ridges found in all the world's oceans, called mid-ocean ridges) is a dividing line between 2 plates that are moving apart. It is also the spot where Earth's surface is being born.

The mid-ocean ridge system encircles the Earth like the seams in a baseball (Figure 3.1), and at these locations, plates are moving away from each other. The Earth doesn't keep getting bigger and bigger, so elsewhere—at subduction zones, created at convergent boundaries like the one between South America and the Nazca Plate—one plate dives down into the Earth when it runs into another one. Subduction zones are also volcanic regions (see Chapter 2), and because of their relative accessibility, are better understood than are Earth's mid-ocean ridges. As the plates diverge at mid-ocean ridges, the pressure on the underlying mantle is released, and it begins to melt. This melt is generally less dense than the surrounding material (hot things rise because they are less dense than the colder material surrounding them) and it works its way to the surface to erupt, making a lava flow at the mid-ocean ridge. This simplistic picture glosses over many fascinating details of mid-ocean ridge volcanism—not the least of which is a description of just how we collect data from mid-ocean ridges.

## 3.3   DATA TYPES AND COLLECTION AT MID-OCEAN RIDGES

All detailed data for volcanism at mid-ocean ridges must be collected from a ship. This means sea-time: sometimes 2–3 months living on a ship, seeing the same 30–50 people every day, and being unable to call your loved ones unless you're willing to pay $10–$50 per minute. Of course, it also means 24 hours of data collection every day, which is thrilling and often surprising. Collecting information to answer specific questions about how the Earth works is why we're out there, after all.

When collecting data at sea there is a necessary trade-off between resolution and coverage, similar to data collection for another planet (see Chapters 4 and 7): we can cover a lot of ground in not very much detail, or we can closely examine a very small region of the sea floor (or another planet). SeaBeam and SeaMARC bathymetry are the lowest resolution data sets (generally ~150 m/pixel for most mid-ocean ridges) with the greatest areal coverage. These sonar systems send a sound signal from the ship to the ocean floor, and record the amount of time it takes for the signal to return to the ship. Using what we know about the velocity of sound in the ocean, this travel-time can be converted to a distance. These distance points can be collated into a bathymetric (or topographic) map of the sea floor. How detailed the map is depends on the water depth and the sound frequency used. Typically, these sonar systems look at a piece of ocean floor that's about 20–100 m$^2$ and gives it one value for "elevation". Thus, these systems can't see anything in great detail, but by slowly driving the ship around the ocean, we can collect a lot of information about several hundred square kilometers of ocean floor. These systems are often used as recon-naissance: they give us an idea of where the mid-ocean ridge is, and where the volcanoes within it might be.

There are higher resolution sonar systems that can be dragged behind the ship attached to a cable. By dragging them, the instrument is much closer to the ocean floor, and this increases the signal resolution. An example is a system called the Deep Submergence Laboratory 120-kHz sonar (*DSL-120*), designed and operated by Woods Hole Oceanographic Institution (Figure 3.3). This instrument is typically dragged about 100 m off the sea floor, allowing features as small as 2–5 m to be resolved. The *DSL-120* is often used prior to sending a manned submarine or a remote-operated vehicle (ROV) (Figure 3.4) to help pinpoint precise landing sites and targets for these even higher resolution instruments.

Tow-cameras are instruments dragged behind the ship a few meters off the sea floor. They commonly collect both continuous video and digital still images (photo-graphs taken every few seconds as the instrument is dragged along). The field of view from tow-cameras is typically 5–10 m, giving highly detailed information about a postage-stamp-sized bit of the sea floor. By mosaicking together adjacent images,

**Figure 3.3.** The *DSL-120* sonar instrument, being lowered into the ocean to begin a survey.
Image courtesy of Woods Hole Oceanographic Institution, National Deep Submergence Facility, and the National
Science Foundation, NOAA and ONR.

and incorporating video data, a swath of images can be created, which gives a continuous (but still narrow) view of the mid-ocean ridge study area.

ROVs and submarines are the ultimate high-resolution data-collecting devices. With superb cameras, often very high-resolution sonar instruments, and even people on board, these instruments can collect data on objects and animals that are less than a centimeter across. The trade-off, however, is that only a few linear kilometers of sea floor (commonly 1–3 km) can be covered during a single submarine dive. The submarine used most often for scientific research in the U.S.A. is the Deep Submergence Vehicle (DSV) *Alvin*, which is owned by the US Navy but operated by Woods Hole Oceanographic Institution (see http://www.whoi.edu/marops/vehicles/alvin/index.html).

Only a few countries own and operate DSVs for scientific research. The Japanese have a research submarine, the *Shinkai 6500*, that can go as deep as 6,500 m, but its batteries take 24 hours to recharge (see http://www.jamstec.go.jp/jamstec-e/shinkai/ for mid-ocean ridge information). That means that the *Shinkai 6500* can only dive every other day; *Alvin*, in contrast, dives every day that weather allows. The *Shinkai 6500* also holds 3 people per dive: usually one pilot and two scientists. France owns the *Nautile* (see http://www.ifremer.

**Figure 3.4.** (a) The *Jason/Medea II* ROV. Like all ROVs, *Jason* is tethered to the mother ship. *Jason* is the large piece with the light-colored roof being lowered into the ocean. *Medea II* is a smaller instrument, hiding behind small doors on the front, that can be released from *Jason* (although *Medea II* remains tethered to *Jason*) for mid-ocean ridge detailed surveys, or to reach places where *Jason* can't go. This ROV was initially developed by Dr. Bob Ballard to investigate the wreck of *RMS Titanic*. (b) The ROV *Ropos* (Canada). *Ropos* is tethered to the mother ship with a fiber-optic cable, and has been used extensively in the exploration of the Juan de Fuca and Gorda Ridges.

(a) Image courtesy of W.W. Chadwick, Jr. (b) Image courtesy of Woods Hole Oceanographic Institution, National Deep Submergence Facility, and the National Science Foundation, NOAA and ONR.

**Figure 3.5.** The DSV *Alvin*. Note that the entire length of the submersible is less than 6 m (approximately 18 ft).
Image courtesy of Woods Hole Oceanographic Institution, National Deep Submergence Facility, and the National Science Foundation, NOAA and ONR.

fr/fleet/systemes_sm/engins/nautile.htm), a submersible capable of carrying three people to depths of 6,000 m. Although I have never been on the *Nautile*, I am told that the main difference between it and *Alvin* is the quality of the lunches packed on board for the divers! Russia owns two identical DSVs: the *Mir-1* and the *Mir-2* (see http://www.sio.rssi.ru/index_en.htm), operated by the Shirshov Institute of Oceanology. Both are capable of diving to depths of 6,000 m. The *Mirs* are unique in the world of manned submersible research because they carry sufficient battery power to enable the divers to stay submerged for 17–20 hours. In contrast, the *Nautile* and the *Alvin* are designed for 8-hour dives. Unlike the *Alvin* and *Nautile*, the *Mirs* carry two pilots and one scientist.

I have been fortunate to experience several dives in the DSV *Alvin* along the mid-ocean ridge in the Pacific Ocean, called the East Pacific Rise (Figure 3.5). With each dive, we learn something new and fascinating about our Earth.

## 3.4 DIVING IN THE DSV *ALVIN*

Diving in *Alvin* is a fantastic experience—as close as one can get to being an astronaut, I imagine, without actually leaving Earth. (For those interested in *Alvin*'s history, please see *Water Baby* (Kaharl, 1990)). *Alvin* is safety certified to dive as deep as 4,500 m, but the average depth of mid-ocean ridges is about 2,500 m; the deepest I've ever gone is almost 3,000 m. *Alvin* is a very small submarine: from tip to tail it is 5.5 m (18 ft) long, but most of that represents the propulsion system and the substantial battery packs. *Alvin* is untethered to any mother ship when it dives, so all its power comes from 2 batteries that are recharged every night. The three divers and all dry equipment are crammed into a titanium sphere that is only 2.6 m (8 ft) in diameter. It's a tight fit, to say the least. *Alvin* is pressurized, so we don't need to bring along scuba gear, although the submarine contains $CO_2$-scrubbers and extra oxygen tanks, just in case. (During your safety check, the pilots will remind you, with a mischievous grin, that *Alvin* has enough oxygen on board for 3 people to last 3 days—or for 1 person to last 9 days.)

To minimize weight (and therefore maximize both battery life and the amount of samples that can be returned) there is no heater in the submarine, nor is there a bathroom. The ambient temperature on the deep sea floor is around 1°C, and it gets chilly in *Alvin* after an hour or so. We compensate by wearing warm clothes (no shoes, though, as shoes track in dirt) and bringing wool blankets with us. Instead of a bathroom, each passenger is equipped with their personal Human Endurance Range Extender (HERE) bottle. This little red bottle with a funnel was designed for men to use, and is almost impossible for a woman to use in such close quarters. It can be done, of course, but it disrupts the entire dive while "privacy screens" (wool blankets held up by the other scientist and the pilot) are constructed. I deal with the issue by not eating after noon and not drinking after 6:00 p.m. the day before a dive.

On a dive day, we're supposed to be in the submarine by 8:00 a.m., and the sub is launched by 8:30. The sub is launched with about 500 pounds of scrap iron hanging from it, which makes the sub sink (again, saving battery power). It takes about 60–90 minutes to reach the mid-ocean ridge. At the bottom, half the scrap iron is dropped and enough seawater taken in the ballast tanks to make the sub neutrally buoyant. We then have about 4–6 hours of "bottom time", cruising along at 1–2 knots, before we drop the remaining scrap iron and float to the surface. Contracts with the submarine pilots and engineers stipulate that we must be back on deck by 5:30 p.m., or they get paid overtime salary. The scrap iron (around 500 pounds of it for each dive) remains on the bottom until it rusts away (which will take centuries)— and serves as a reminder to future divers that someone has visited that place before. We always try to "land" and "take off" at least 100 m from any place that might have biologic activity so as to not contaminate it with *Alvin*'s iron.

In my investigations of mid-ocean ridges, I rely most heavily on data collected by *Alvin*, ROVs, and towed instruments (such as the *DSL-120*). This is because I'm interested in the smaller scale features that can be found at mid-ocean ridges, such as individual lava flows. These are the products of single eruptions and therefore form the fundamental building blocks for oceanic crust.

## 3.5    MID-OCEAN RIDGE CLASSIFICATION

Earth's tectonic plates do not all move at the same rate, and most of the characteristics of mid-ocean ridges can be related to the rate at which the plates are diverging. This is called the spreading rate, and here they are presented as full spreading rates:

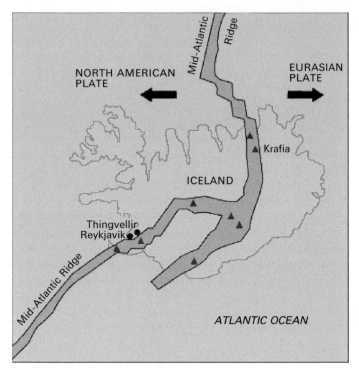

**Figure 3.6.** The Mid-Atlantic Ridge cutting through Iceland. The full spreading rate here is about 2.5 cm/yr. If you were to stand on the North American Plate, it would appear that the Eurasion Plate was moving away from you at the rate of 2.5 cm/yr, and this is the *full spreading rate*. If you were to stand in the ridge—the dark swath in the figure—it would appear that the Eurasian Plate was moving away from you at the rate of 1.25 cm/yr, and this is called the *half spreading rate*.
Image courtesy of the USGS.

that is, how fast plate B would appear to be moving away from you if you were standing on the other side of the ridge on plate A (Figure 3.6).

Mid-ocean ridges are classified as slow, intermediate, and fast—with the superlative categories of "ultraslow" and "ultrafast" being used in some extreme cases (see Table 3.1). The type examples for slow, intermediate, and fast spreading rates are the Mid-Atlantic Ridge, the Juan de Fuca Ridge, and the East Pacific Rise near 10°N, respectively. The volcanic behavior of a mid-ocean ridge can be directly related to the spreading rate.

Some characteristics, however, are common to all mid-ocean ridges. First, all mid-ocean ridges erupt basalt almost exclusively. This is the same kind of lava found on the volcanic islands of Hawaii (see Chapter 2) and is probably the most abundant lava type on Mars (see Chapter 7) and Venus (see Chapter 4). This is the type of melt that is produced when Earth's mantle is heated. Andesite is a lava type that forms when the initial mantle melt is altered in some way—typically by partially freezing on

**Table 3.1.** Spreading rates for slow through ultrafast mid-ocean ridges.

| Ridge | Full spreading rate (cm/yr) | Classification |
|---|---|---|
| Southeast Indian | <1.0 | Ultraslow |
| Mid-Atlantic | 2.5 | Slow |
| Juan de Fuca | 6.0–9.0 | Intermediate |
| East Pacific Rise (9–10°N) | 9.0–11.0 | Fast |
| East Pacific Rise (17.5°S) | 16–20 | Ultrafast |

its way to the surface. Andesite has been found on the Galapagos Ridge, but it is extremely rare in mid-ocean ridge environments—basalt dominates mid-ocean ridge volcanism. Second, all mid-ocean ridges are higher than the surrounding terrain. Just how much higher depends on spreading rate, but all mid-ocean ridges are essentially submarine mountain chains. Third, a valley, trough or depression is commonly found at the summit of the mid-ocean ridge. The dimensions and genesis of this trough differ from place to place, but most ridges have one.

## 3.6   SLOW-SPREADING CENTERS: THE MID-ATLANTIC RIDGE

Most students learn in their introductory geology classes that the ocean crust is made up of pillow basalts (Figure 3.7(a)). This lava morphology only forms under water, but it does not in fact make up the majority of the oceanic crust. Instead, lobate lavas (Figure 3.7(b)) are most abundant. The reason most people still think that the sea floor is composed of pillow lavas is that the very first mid-ocean ridge to be studied (in the late 1970s) using photographic techniques was the Mid-Atlantic Ridge, and it just so happens that the Mid-Atlantic Ridge *is* dominated by pillow basalts. Work done by my graduate advisor, Jonathan Fink, his colleague Ross Griffiths, and myself (Griffiths and Fink, 1992; Gregg and Fink, 1992) demonstrated that pillow basalts form when the lava comes out of the ground relatively slowly, and/or has a relatively high viscosity. We believe that the pillow basalts at the Mid-Atlantic Ridge are a product of a low effusion rate, and cool, crystal-rich lavas erupting there.

Slow spreading centers (full spreading rate ≤2 cm/year) have remarkable bathymetry. They are characterized by a 2–10 km wide rift valley at the summit of a ridge that may be 4–6 km above the surrounding ocean floor. The rift valley floor contains large shield volcanoes, similar to (but smaller than) those we find on the volcanic islands of Hawaii (see Chapter 2) (Figure 3.8; see color section); Debbie Smith at Woods Hole Oceanographic Institution has spent a great deal of her career looking at these features (Smith *et al.*, 1995; Smith and Cann, 1999). Individual volcanoes tend to be elongate because they erupted from long cracks, or fissures, on the rift valley floor. These fissures are created by the stresses of the diverging plates. Faults, formed by tensional stresses of diverging plates, cover the walls of the rift valley, and

(a)

(b)

**Figure 3.7.** (a) A pillow lava erupted on the Juan de Fuca Ridge. The pillow is about 2 m across. The grayish material is sediment composed largely of sulfur, associated with a recent eruption. This image was taken from a video collected by the ROV *Ropos*. (b) A field of lobate lavas erupted onto the Juan de Fuca Ridge. This image was taken from a tow-camera, dragged about 5 m above the sea floor, and is about 4 m across. The lighter material between the black basalt lobes is mostly a biologic ooze composed of dead animals/plants that have descended from the ocean surface. There is also a small component of volcanic sulfur.

extend out onto the surrounding ocean floor; occasionally a small shield volcano may form on top of one of these wall faults.

The shield volcanoes in the rift valley floor may be as long as 5 km and as tall as 2 km. No one has ever observed a volcanic eruption at a slow-spreading center, although instruments known as Ocean Bottom Seismometers have been placed to listen for the sounds of magma moving beneath the surface. These instruments record sounds made at a much lower frequency than human ears can detect, and are essentially the same as magnitude 2.0–3.0 earthquakes on land. These low-magnitude earthquakes are seldom even noticed by people. A few promising sounds have been recorded, but we still know very little about the individual volcanic eruptions at slow-spreading centers. We suspect that they are very infrequent—one every 1,000–10,000 years per kilometer of ridge or so—but are probably long-lived. Observations based on the shapes of the rift valley volcanoes lead us to believe that they are constructed by only a few individual eruptions. By comparing estimated lava effusion rates with volumes of volcanoes, we can calculate an eruption duration of years to decades (and, in some cases, centuries).

Thus, eruptions at slow-spreading centers are large-volume (probably $10^8$–$10^{10}$ m$^3$), long-duration events that tend to erupt cool, crystal-rich lavas.

## 3.7    FAST-SPREADING CENTERS: THE EAST PACIFIC RISE NEAR 10°N

In contrast to the slow-spreading Mid-Atlantic Ridge, the East Pacific Rise near 10°N spreads at 9–11 cm/yr. It is characterized by a broad (5–10 km), low (<200 m) rise that is capped in most places by a trough that is 40–200 m wide and 6–20 m deep. This trough, termed the "axial summit collapse trough" or ASCT by Dan Fornari and others (Fornari *et al.*, 1998) is morphologically analogous to the much larger rift valley found at the slow-spreading Mid-Atlantic Ridge. The main difference, however, is in their formation. The Mid-Atlantic Ridge rift valley is produced primarily by tectonic forces (i.e., the pulling apart of the plates creates tensional faulting; the rift valley is essentially a fault-bounded valley known as a graben). In contrast, the axial summit collapse trough along the northern East Pacific Rise is, in most places, created by near-surface volcanic processes. Although there is still lively debate on this topic (see, e.g., Chadwick and Embley, 1998), I believe that the axial summit collapse trough is formed through repeated, frequent, small-volume volcanic eruptions and subsequent drainback of the lava into the fissure and collapse of the overlying lava crust.

We know much more about volcanic eruptions at the northern East Pacific Rise than we do about those at the slow-spreading Mid-Atlantic Ridge, largely because they are more frequent at the East Pacific Rise (Perfit and Chadwick, 1998). Furthermore, in 1991, science and serendipity collided, as they often do. A cruise to the northern East Pacific Rise in April of that year had been launched to investigate the behavior of hydrothermal waters associated with the northern East Pacific Rise. Hydrothermal vents have been found on all mid-ocean ridges studied to date, and are a natural part of the interaction of tons of ocean water with hot magma and/or rocks that exist beneath all mid-ocean ridges. When *Alvin* reached the sea floor that fateful April, however, the scientists didn't find what they expected—instead, they found fresh, black, glassy lava; dead animal communities (not yet destroyed by scavengers) over-run by lava (one such site was named "Tubeworm Barbeque"); and hot water pouring directly out of holes and cracks in the lava surface rather than being confined to a few precise locations. They interpreted this to mean that they were either watching an eruption in action, or had arrived very shortly after the eruption stopped (Haymon *et al.*, 1993). Subsequent radiometric dating of the fresh lava revealed that they had arrived 2–4 weeks after the eruption (Rubin *et al.*, 1994).

This gave volcanologists an unprecedented opportunity to investigate mid-ocean ridge eruptions. On land, if a volcanologist wants to understand the past, present, and future behavior of a volcano, the first step is to determine when eruptions happened in the past, and what they were like. Prior to this 1991 East Pacific Rise eruption, that couldn't be done for mid-ocean ridges. But here was a fresh lava that was easily distinguished from the older lavas beneath it (so it could be readily mapped using *Alvin*), and a pretty precise timing of the event.

I relied heavily on information collected about this eruption to construct our current understanding of volcanic eruptions at fast-spreading centers. It seems that eruptions at the northern East Pacific Rise are small-volume, the order of $10^6 \, m^3$. They are frequent, with an eruption occurring once every 5 years or so per kilometer of ridge. Finally, they are short-lived. The lava erupted in 1991 was probably emplaced in less than a day, with most of the lava coming out of the 8 km long fissure in just 2–4 hours or so.

These observations and interpretations, when coupled with seismic studies of the state of the mantle beneath the East Pacific Rise, suggest that there is always a ready supply of magma just beneath the surface, waiting to erupt. The magma resides in a long (10–20 km), narrow (<2 km wide), shallow (10–100 m deep) chamber beneath the ridge axis. The high-spreading rate places the brittle lid above the magma chamber in a constant state of extensional stress, and so the lid cracks frequently. Magma appears to be continually pouring into the chamber from sources deeper in the mantle, and this places the chamber in an almost constant state of being over-pressurized. When the brittle lid cracks, the magma leaps out quickly, but just enough comes out to relieve the pressure. Once the pressure is relieved, the eruption ends, and another eruption can't occur until the magma chamber refills sufficiently. Thus, the northern East Pacific Rise goes through a volcanic–tectonic cycle. Arbitrarily, we consider the cycle to begin with an eruption. The tectonic stresses and magmatic stresses have been temporarily relieved. With time, the tectonic stresses continue to build as the plates are tugged apart, causing minor near-surface cracking and disruption of the young lava flow. Meanwhile, the magma chamber continues to fill and becomes pressurized. When the magma chamber is sufficiently pressurized, a small tectonic crack will become a conduit for the next volcanic eruption.

## 3.8  INTERMEDIATE-SPREADING CENTERS

We know the most about intermediate-spreading centers—specifically, the Juan de Fuca and Gorda Ridges, which have full spreading rates of 6–9 cm/yr. The Juan de Fuca and Gorda Ridges are located just a few hundred kilometers from the coasts of Oregon and Washington in the northwestern U.S.A., and thus can be reached by ship in just a day or two, depending on the weather. These mid-ocean ridges are so close to land that instruments can be placed on these ridges and wired so that they can be monitored in real time. Elsewhere on the mid-ocean ridge system, instruments are deployed and left alone to record data for months or years. The instruments are eventually retrieved, and the data downloaded. There are obvious drawbacks to this method of data collection, but the remoteness of most mid-ocean ridges makes it the only possible mechanism to use. On the Juan de Fuca Ridge, data can be transmitted to receiving stations as they are collected. For example, the US Navy installed a network of hydrophones (underwater microphones) initially designed to listen for enemy submarines up and down the west coast of the U.S.A.. In the mid-1990s, the Navy was no longer terribly concerned about the threat of invading submarines, nor did it have the funds to support the hydrophone array. The Navy then turned over the operation of these hydrophones to NOAA. NOAA has offices in Newport, Oregon, and some of the scientists there are deeply interested in Mid-ocean ridge eruptions. These scientists arranged with NOAA to have the real-time hydrophone data transmitted directly into their Newport facility. Four days after this transfer of responsibility, the scientists at NOAA were rewarded by being the first people to hear a mid-ocean ridge eruption (see `http://www.pmel.noaa.gov/vents/acoustics/ sounds/tremor.html` to listen to the sounds of an underwater eruption). Since that day in June, 1994, the hydrophone array has detected 5 other eruptions. In most cases, a research ship carrying tow-cameras, ROVs, and other instruments, has been able to reach the eruption site within a few days or weeks, adding enormously to our understanding of mid-ocean ridge processes.

Intermediate-spreading centers share some characteristics with both fast and slow-spreading centers, depending on the location. For example, some regions of the Juan de Fuca Ridge (e.g., near 50°10′ N, part of the Explorer Ridge) look just like the slow-spreading Mid-Atlantic Ridge complete with a deep, broad rift valley. Near 44°39′ N, the Juan de Fuca Ridge looks remarkably similar to the fast-spreading East Pacific Rise. In other words, the morphology of the mid-ocean ridge is defined by the relative roles of volcanism and tectonics. At slow-spreading centers, there is a limited magma supply over the short term, but the tension caused by plates moving apart is constant. Therefore, lots of faults form, before enough magma has built up to generate a large eruption. In contrast, the fast-spreading centers have an abundant magma supply. The constant tectonic tension forms cracks in the magma chamber lid that allow the ready and waiting magma to escape frequently.

At intermediate-spreading centers, eruption volumes are smaller than those at slow-spreading centers but larger than those at fast-spreading centers (probably around $10^8$–$10^9$ m$^3$). Hydrophone data suggests that there is one eruption every 5–10 years per kilometer of ridge, and that the eruption duration is over a period of days to weeks.

## 3.9  HYDROTHERMAL VENTING AT MID-OCEAN RIDGES

Although not my specialty, hydrothermal venting at mid-ocean ridges plays a vital role in cooling the Earth, and it is impossible to study mid-ocean ridge volcanic processes without having at least a working knowledge of the hydrothermal processes there.

Hydrothermal venting has been found along every mid-ocean ridge that has been explored, although the styles of venting differ from place to place. What they all share is a link to the Earth's internal heat and access to lots of water. The Earth's mantle is very close to the surface at mid-ocean ridges, and magma with temperatures of almost 1,200°C lurk only a few hundred meters beneath an ultrafast spreading ridge axis. Ocean water percolates down through holes and cracks in the oceanic crust until it becomes heated by the hot (and, in some cases, molten) material at depth. The ocean water is heated to amazingly high temperatures (higher than 540°C) because the great pressure of the overlying rock and water prevent the water from boiling. The hot ocean water rises up through the cracks and holes in the basaltic crust, leaching out minerals from the rock as it goes—just like hot water can corrode the pipes in your house. Once the hydrothermal water reaches the ocean floor, it has cooled a little but may still be as hot as 450°C, and it carries a heavy load of dissolved minerals. The hydrothermal water comes out of the ocean floor and comes into contact with the cold ($\sim$1°C) ambient seawater. This rapidly chills the hydrothermal water, and the dissolved minerals immediately precipitate—come out of solution—creating what looks like a plume of smoke but what is in reality a plume of hot, chemical-laden water. The hottest vents are called "black smokers" because of their appearance. In other places, hydrothermal waters may seep out of the ocean floor with temperatures of 20°C–25°C and can support large colonies of animal life. Among the most famous are tube worms (Figure 3.9). The base of the food chain here is sulfur—life at mid-ocean ridges is chemosynthetic rather than the photosynthetic plant life elsewhere on Earth. Basalts contain large amounts of dissolved sulfur, and this is leached out and carried to the ocean floor by hydrothermal activity. Particular strains of bacteria (some thought to be so ancient and primitive that they are called archeobacteria) have evolved to consume the sulfur to survive. All other life at mid-ocean ridges either consumes these bacteria directly or has evolved a symbiotic relationship with them to enable them to live in this realm without sunlight.

It has been proposed that similar conditions—lots of water and a source of internal planetary heat—may exist on Europa. If so, it is possible that bacterial life has developed there, too.

**Figure 3.9.** Tubeworms from the Galapagos Rift.

## 3.10   SUMMARY

The mid-ocean ridge system is the most volcanically active region on Earth, and yet we have examined <1% of this system in any detail. Mid-ocean ridges are where new oceanic crust is being born, and its likely that somewhere on Earth, there is a mid-ocean ridge eruption every day. There is a strong correlation between volcanic style, eruption volume, frequency, duration, and spreading rate: fast-spreading centers have low-volume, frequent and short-duration eruptions whereas slow-spreading centers have large-volume, infrequent and long-duration eruptions. Earth is the only planet in the solar system with plate tectonics and a global system of spreading centers. Why should this be so? Venus is about the same size as Earth and located about the same distance from the Sun (see Chapter 4). Although there has clearly been a lot of tectonic and volcanic activity on Venus, it is not organized in the same way that plate tectonics organizes these things. Similarly, Mars also has volcanism and tectonics, but it does not have plate tectonics. It has been argued that Mars is too small to contain sufficient heat to drive plate tectonics for any length of time; some evidence from Mars Global Surveyor (Acuna *et al.*, 2001) suggests that Mars may have had Earth-like plate tectonics for a few hundred million years early in its existence. Europa, the Moon, Mercury, all the solid planetary bodies in the solar system, exhibit some style of tectonism (defined as vertical or lateral movements of the lithosphere or crust) and volcanism—but only on Earth have these arranged themselves as "plate tectonics". Intriguingly, Earth is also the only planet in the solar system to have extant life (as far as we know) and liquid water stable on its surface. There may be a connection between life, water, and plate tectonics. For example, when two plates composed of oceanic crust collide on

Earth, one of them must subduct. The one that does is typically older; it is covered with more sediment—which is composed mostly of the remains of dead plankton—and is colder, so it's heavier than the younger plate. Without liquid water cooling the oceanic plate, and without abundant life raining down sediment on the ocean floor, it's not clear if one plate would necessarily subduct: perhaps they would merely crumple up, forming ridge belts.

The connection between life, liquid water, and plate tectonics is not a straight-forward one. But it is clear that further investigations of the Earth's mid-ocean ridges will continue to enlighten us about the inner workings of our own planet—and those of other planets as well.

## 3.11  REFERENCES

Acuna, M.H., Connerney, J.E.P., Wasilewski, P., Lin, R.P., Mitchell, D.L., Anderson, K.A., Carlson, C.W., McFadden, J., Reme, H., Mazelle, C., *et al.* (2001) Magnetic field of Mars: Summary of results from the aerobraking and mapping orbits. *J. Geophys. Res.,* **106**, 23403–23417.

Chadwick, W.W. Jr. and Embley, R.W. (1998) Graben formation associated with recent dike intrusions and volcanic eruptions on the mid-ocean ridge. *J. Geophys. Res.,* **103**, 9807–9825.

Dietz, R.S. (1961) Continent and ocean basin evolution by spreading of the sea floor. *Nature,* **90**, 854–857.

Ewing, M. and Heezen, B.C. (1960) Continuity of the mid-oceanic ridge and rift valley in the southwestern indian ocean confirmed. *Science,* **131**, 1677–1679.

Fornari, D.J., Haymon, R.M., Perfit, M.R., Gregg, T.K.P., and Edwards, M.H. (1998) Axial summit trough of the East Pacific Rise 9°–10°N: Geological characteristics and evolution of the axial zone on fast spreading mid-ocean ridges. *J. Geophys. Res.,* **103**, 9827–9855.

Gregg, T.K.P. and Fink, J.H. (1995) Quantification of submarine lava-flow morphology through analog experiments. *Geology,* **23**, 73–76.

Griffiths, R.W. and Fink, J.H. (1992) Solidification and morphology of submarine lavas; A dependence on extrusion rate. *J. Geophys. Res.* **97**, 19729–19737.

Haymon, R.M., Fornari, D.J., Von Damm, K.L., Lilley, M.D., Perfit, M.R., Edmond, J.M., Shanks III, W.C., Lutz, R.A., Grebmeier, J.M., Carbotte, S., *et al.* (1993) Volcanic eruption of the mid-ocean ridge along the East Pacific Rise crest at 9° 45–52′N: Direct submersible observations of sea floor phenomena associated with an eruption event in April, 1991. *Earth Planet. Sci. Lett.,* **119**, 85–101.

Heezen, B.C., Tharp, M., and Ewing, M. (1959) The floors of the oceans. I: The North Atlantic. Geol. Soc. Am. Spec. Paper 65.

Hess, H.H. (1962) History of the ocean basins. In: A.E.J. Engel and others (eds), *Petrologic Studies: A volume in Honor of Al Buddinton.* Geological Society of America, Boulder, CO, pp. 599–620.

Kaharl, V. (1990) *Water Baby.* Oxford University Press, New York, 356 pp.

Perfit, M.R. and Chadwick, W.W. Jr. (1998) Magmatism at mid-ocean ridges: Constraints from volcanological and geochemical investigations. In: W.R. Buck, P.T. Delany, J.A. Karson, and Y Lagabrielle (eds), *Faulting and Magmatism at Mid-Ocean Ridges.* American Geophysical Union, Washington, DC, pp. 59–115.

Rubin, K.H., Macdougall, J.D. and Perfit, M.R. (1994) $^{210}$Po–$^{210}$Pb dating of recent volcanic eruptions on the sea floor. *Nature*, **368**, 841 844.

Smith, D.K. and Cann, J.R. (1999) Constructing the upper crust of the Mid-Atlantic Ridge: A reinterpretation based on the Puna Ridge, Kilauea volcano. *J. Geophys. Res.*, **104**, 25379–25399.

Smith, D.K., Cann, J.R., Dougherty, M.E., Lin, J., Spencer, S., MacLeod, C., Keeton, J., McAllister, E., Brooks, B., Pascoe, R., and Robertson, W. (1995) Mid-Atlantic Ridge volcanism from deep-towed side-scan sonar images, 25°–29°N. *J. Volcanol. Geotherm. Res.*, **67**, 233–262.

## Websites

Woods Hole Oceanographic Institution and *Alvin:* http://www.whoi.edu/marops/vehicles/alvin/index.html

*Shinkai 6500*: http://www.jamstec.go.jp/jamstec-e/shinkai/

IFREMER and Nautile: http://www.ifremer.fr/fleet/systemes_sm/engins/nautile.htm

Shirshov School of Oceanology: http://www.sio.rssi.ru/index_en.htm

NOAA/PMEL VENTS program: http://www.pmel.noaa.gov/vents/

NEMO Explorer program: http://www.pmel.noaa.gov/vents/nemo/explorer.html

Submarine accoustic monitoring: http://www.pmel.noaa.gov/vents/acoustics/sounds/tremor.html

# 4

# Earth's evil twin: The volcanic world of Venus

*Ellen Stofan* (Proxemy Research)

*I am Ellen Stofan, a planetary geologist who studies volcanoes on Venus, Earth, and Mars. When I was young, I wanted to be an archaeologist, but attending a geology field trip with my science teacher mother convinced me that geology was fun—you got to go on hikes and pick up rocks for a job! My father worked for NASA, so it was natural for me to combine my interest in geology with my love of space exploration to study the surfaces of planets. I have an undergraduate degree in geology from the College of William and Mary in Virginia, and MSc and PhD degrees in geological sciences from Brown University in Rhode Island. I worked for the Jet Propulsion Laboratory in Pasadena, California, for about ten years. I was the Deputy Project Scientist on the Magellan mission to Venus, the Experiment Scientist on the Spaceborne Imaging Radar project (SIR-C/X-SAR) that flew on the space shuttle twice in 1994, and the Chief Scientist for the Exploration for the New Millennium program. Since leaving JPL in 2000 to spend more time with my three children, I work part time for a company called Proxemy Research and as a Research Fellow at University College London. I conduct research on volcanic features on Venus, Mars, and Earth, and do fieldwork in California, Oregon, Hawaii, and Sicily.*

Venus is a familiar object in the sky, known commonly as the Evening Star. However, although it is usually one of the brightest objects in the night or early morning sky, it is a planet, at 0.723 AU ($108.2 \times 10^6$ km from the Sun) (Figure 4.1; see color section). Venus has no satellite, and orbits very slowly (one orbital period is 243 days) retrograde, or in the opposite direction to Earth's rotation (Cattermole, 1994). Venus does not violate any laws of physics by rotating in an opposite direction to that of Earth, but it has also been proposed that Venus rotates "backwards" from being hit early in its history by a large body.

Venus is frequently called Earth's twin, with nearly the same mass and radius (Venus' equatorial radius = 6,052 km, Earth's equatorial radius = 6,378 km).

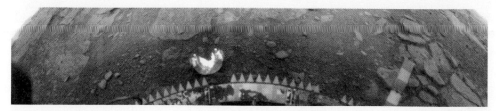

**Figure 4.2.** This image of the Venusian surface was taken by the Russian Venera 13 lander, which landed on the surface of Venus on 1 March, 1982. The landing site was at approximately 7°S latitude, 305°E longitude, near Phoebe Regio. The platy surface is similar to a terrestrial lava flow, and the lander measured a basaltic composition of the surface rocks.

Also, its position so close to Earth suggests that it should be made of approximately the same material as Earth according to most planetary accretion models. However, the planet's surface is at approximately 750°C (900°F) and 90 bars pressure, due to the greenhouse effect produced by its perpetual cloud cover. These clouds are composed mostly of carbon dioxide gas, which on Earth is locked up in rocks such as limestone. The atmosphere contains corrosive droplets of hydrochloric acid, which make exploring the planet difficult. Liquid water cannot exist on the surface at those temperatures, nor life as we know it. Why is this planet so like Earth in some ways, and yet has such a hellish surface? Earth and Venus both formed about 4.5 billion years ago from similar materials in a relatively similar place in the solar system, but Earth evolved into a hospitable blue planet, while Venus became Earth's evil twin (Grinspoon, 1998).   ·

The surface of Venus was first studied in detail by the then Soviet Union, who landed a series of probes on the surface, and mapped part of the planet with an orbiting radar (Surkov, 1983; Barsukov *et al.*, 1984). Radar waves can penetrate through the thick cloud cover on Venus, unlike shorter wavelength visible light, which is mostly reflected back or scattered away by cloud droplets. The radar waves penetrate to the surface and are reflected back to the radar, returning an image that is dominated by changes in surface roughness. The Soviet radar missions showed a surface with many volcanoes and strongly deformed regions named tessera (Basilevsky *et al.*, 1986). The tessera terrain is located in highland regions, and has been suggested to be analogous to granitic (highly silicic) terrestrial continents.

The Soviet lander missions returned visible images that showed a surface that looked more or less like the lava flows in Hawaii, with rock compositions similar to basalt (Surkov, 1983) (Figure 4.2). Some of the landing sites had soil surrounding the rocks (Florensky *et al.*, 1983), indicating that some erosion has taken place on the surface, despite the lack of water that is the main "erosive" agent on Earth.

The U.S.A. also sent missions to Venus, starting with Mariner 2 in 1962, then Mariner 5 in 1967, Mariner 10 in 1975, and Pioneer Venus in 1978. Pioneer Venus returned information on the composition of the atmosphere and the topography, or surface heights, of the planet (Pettengill *et al.*, 1980; Masursky *et al.*, 1980). The whole surface of Venus was revealed in detail by radar images from the NASA Magellan spacecraft in the early 1990s, at about 120-m resolution (the Soviet radar images had a resolution of about 10 km; Pioneer Venus topography about 100 km) (Saunders *et al.*, 1991; Saunders *et al.*, 1992). Nearly the entire surface of the planet was mapped, along with higher resolution measurements of the surface topography, and measurements of the planet's gravity field. These data provided a better global view of Venus than we have of the Earth, due to the fact that most of the Earth's surface is covered by water. The data of Venus showed that Earth's twin has undergone a very different geologic history from that of the Earth, with no evidence seen for plate tectonics. Perhaps the high surface temperatures of Venus and its lack of liquid water are the key to the differences between the two planets.

What we did see on Venus was many superposed volcanic plains units that covered 100s of kilometers, providing evidence for multiple episodes of vast outpourings of lava (Head *et al.*, 1992). Much of the surface is covered by volcanic features of one type or another, from volcanoes to floods of lava to channels carved by lava. Thousands of volcanoes have been identified on the surface of Venus, some of them the most unusual volcanoes in the solar system.

## 4.1  TYPES OF VOLCANIC FEATURES ON VENUS

Volcanoes on Venus come in all shapes and sizes. The smallest volcanoes are less than 10 km across (for scale, Kilauea volcano in Hawaii is about 18 km across). Small volcanoes come in a variety of shapes, from shallow sloping shields to cones with steeper slopes (Figure 4.3) (Guest *et al.*, 1992). Many of the shields and cones have summit pits, which provide evidence for either collapse into a magma chamber at the summit or small explosive eruptions (large explosive eruptions are unlikely due to the lack of water). Shields are much more common than cones, suggesting that they were formed by relatively fluid lavas that did not allow steep slopes to build up. The cones and shields tend to occur in clusters or groups, and some have flows emanating from their bases. Both old and young shields can be found (Figure 4.4).

Fewer in number but still very common are intermediate volcanoes, those that are 10–100 km across (Figure 4.5) (Crumpler *et al.*, 1997). Radar images of these features show them to be surrounded by flows of varying brightness. The brightness of a lava flow is primarily related to how rough it is, at the scale of the radar wavelength (the Magellan radar wavelength was 12.6 cm). A bright lava flow is therefore blocky, while a dark one has a very smooth surface. Scientists studying the Magellan flows use the radar brightness, the shapes of the lava flows, and their distribution around the volcano to unravel how the volcano formed. Most lava flows around intermediate volcanoes are of intermediate brightness. Comparisons to radar images of lava flows in Hawaii indicate that the Venusian flows were built up of lavas with similar roughness. Despite the higher surface temperatures on Venus, basaltic lavas should form a crust relatively rapidly after eruption (Bridges, 1997) and therefore, should produce lava surface textures similar to those on Earth.

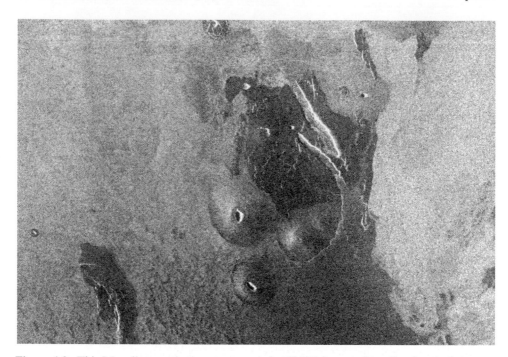

**Figure 4.3.** This Magellan synthetic apperture radar (SAR) image of small volcanic domes on the flank of the volcano Maat Mons is centered about 3.2°N latitude, 194.9°E longitude. The image is 90 km in width and 80 km in length. The bright flows to the east are most likely rough lava flows while the darker flows to the west are probably smoother flows. The dark flows do show some roughness, however, as can be seen by the structure in the flows to the south-west. These dark flows also have some debris that has been deposited on top of the flows. The debris may be fine material deposited from the surrounding plains on top of the flow by wind or it may be ash from the volcano. Small volcanic domes are very common features on the surface of Venus, indicating that there has been much volcanic activity on the surface. Assuming that the central volcanic cone is symmetrical in shape and knowing the length of the cone's side and the incidence angle, radar foreshortening yields a height and slope of 688 m and 8.2°, respectively for the cone. These values are similar to the heights and slopes of some volcanic cones on the Earth.

Large volcanoes on Venus are relatively few in number, and are surrounded by dramatic assemblages of flows (Figure 4.6) (Head *et al.*, 1992; Stofan *et al.*, 2001a). These volcanoes tend to be located on top of broad topographic rises, similar to volcanoes on Earth such as Hawaii that form at hotspots (plumes of rising hot material from deep within the planet, see Chapter 2). These hotspots appear to be randomly located around Venus, with no alignments of volcanoes seen that would indicate plate boundaries. Large volcanoes have different types of summits: some have large calderas while others have sets of radial fractures that likely were produced from diking (the intrusion of magma under the surface). Most large volcanoes are built up of very long flows (>100 km), combined with more

**Figure 4.4.** This Magellan image shows a group of volcanic features on the plains of Venus, with superimposed surface fractures. The image is centered at 9°S latitude, 199°E longitude, and is about 350 km across. Digitate and lobate lava flows emanate from small shields as well as from linear fractures or fissures. Numerous surface fractures and graben (linear valleys) trend north to north-east across the area. While most of the fractures are not buried by the lavas, buried fractures indicate the overlap in time of volcanic and tectonic activity in this region. Resolution of the Magellan data is about 120 m.

abundant shorter flows (Stofan *et al.*, 2001a). Small shields and cones on the flanks of the large volcanoes are similar to those seen at volcanoes on Earth, and indicate that sometimes it is easier for magma to erupt from a lower elevation on the volcano rather than from the volcano's summit. Venus' large volcanoes have much larger volumes than those on Earth, probably because hotspot volcanoes on Earth sit on a plate that is moved off a rising mantle plume (Stofan *et al.*, 1995). Without mobile

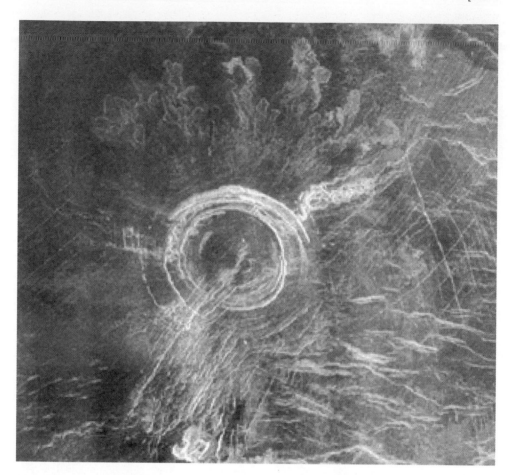

**Figure 4.5.** This Magellan SAR mosaic shows an intermediate-size volcano called Serova Patera. The center of the image is at 20.5°N latitude, 247°E longitude. The volcano is about 80 km across, and consists of a broad caldera surrounded by gently sloping, narrow flanks. The volcano is surrounded by lava flows, and has been fractured by at least two episodes of subsequent tectonic activity. The image is about 200 km across; resolution of the Magellan data is about 120 m.

plates, Venusian volcanoes remain stationary over a plume, and can build up to larger sizes over time. Lava flows on Venus also tend to be longer than flows on Earth (10s of kilometers on Earth against 100s of kilometers on Venus), for reasons that are still unclear.

Another common volcanic feature on Venus is flood lavas, large plains formed by vast outpourings of lava (Figure 4.7) (Lancaster *et al.*, 1995). These volcanic deposits extend for several hundreds of kilometers, and are composed of both radar-bright (rough) and radar-dark (smooth) flows. They have flowed very far, despite the fact that the terrain they form on is extremely flat. These features are

**Figure 4.6.** The volcano Sif Mons is seen in this computer-generated view of the surface of Venus. Magellan SAR data is combined with Magellan altimetry to produce a 3-D view of the surface. Rays, cast in a computer, intersect the surface to create a 3-D perspective view. Simulated color and a digital elevation map (vertically exaggerated) developed by the US Geological Survey (USGS) are used to enhance small-scale structure. The simulated hues are based on color images recorded by the Soviet Venera 13 and 14 spacecraft. The viewpoint is located 360 km (223 miles) north of Sif Mons at a height of 7.5 km (4.7 miles) above the surface. Sif Mons has a diameter of about 350 km and a height of about 2 km. It has very shallow sloping flanks and a 75-km caldera at its summit. Effective resolution of this image is about 225 m. This image was produced at JPL.
See book cover for color version.

also common on Mars (see Chapters 6 and 7), and slightly less common on Earth. On Earth, flood lavas such as the Columbia River Basalts and the Deccan Traps are highly eroded, and we have no examples of flood lavas that have formed in modern times. Therefore, studies of these features on other planets can provide insights into their formation that we could not obtain by only having the terrestrial examples. For example, the flood lavas on Venus are uneroded, and studies of their surface textures have indicated that they were likely to have been emplaced relatively rapidly.

Venus also has several types of volcanic features that differ from those on Earth and other planets. Steep-sided domes, also called pancake domes, are flat-topped, steep-sided features (Figure 4.8) (Pavri *et al.*, 1992). These domes are somewhat

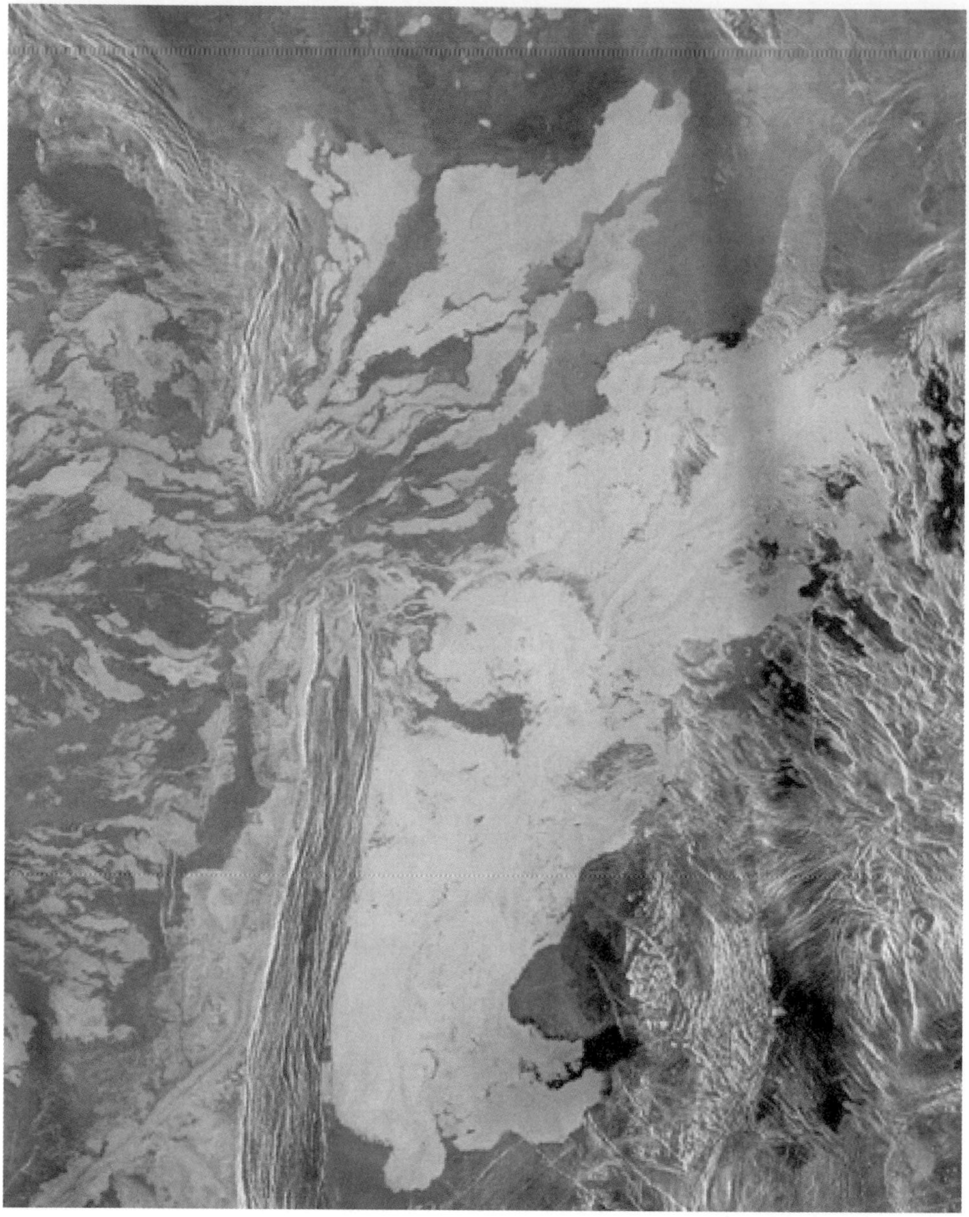

**Figure 4.7.** A large volcanic province is shown in this Magellan image mosaic, composed of radar-bright (rough) and radar-dark (smooth) lava flows that emanated from a caldera 300 km to the west of the scene. The complex of east-trending radar-bright and radar-dark lava flows have breached a north-trending ridge belt (left of center), forming an extensive radar-bright deposit which extends for about 100,000 km² (right side of image). The image mosaic is centered at 47°S latitude, 25°E longitude, and is approximately 550 km wide by 630 km long. The mosaic resolution of this Magellan SAR image is 225 m/pixel.

**Figure 4.8.** This Magellan SAR image is of three flat-topped volcanoes at approximately 9°S latitude, 199°E longitude . At the center of the image is a very circular, large volcano (50 km in diameter) with steep sides and a flat top. This feature overlies a 45 km diameter volcano to the south-west that is cut by numerous fractures. The 25 km diameter volcano to the south-east has scalloped edges, probably the result of multiple landslides along its steep margin. These types of volcanoes are thought to erupt from a point source, and to form from either rapidly flowing or relatively thick and viscous lava flows. Resolution of the Magellan data is about 120 m.

similar to flat-topped domes on Earth that are formed by silicic lavas, such as the Inyo domes in California, but they are much larger, and have smooth rather than blocky surfaces. There is still a fair amount of debate as to how these domes formed, and what kind of lavas they are made of (Stofan *et al.*, 2000). Another type of volcanic dome peculiar to Venus is the scalloped margin domes (Figure 4.9). These features, which were first nicknamed "ticks" due to their odd appearance, are interpreted to be steep-sided domes whose margins have collapsed (Guest *et al.*, 1992).

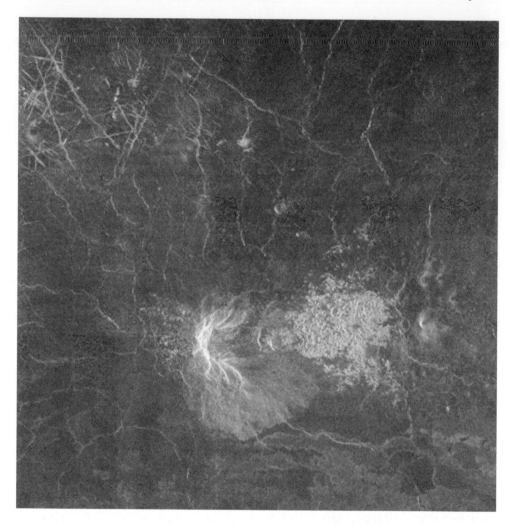

**Figure 4.9.** The volcano at the center of this Magellan SAR image is about 15–20 km in diameter and has an apron of blocky debris to the east and some smaller aprons to the west of the feature. This Magellan image is centered at 55°N latitude, 266°E longitude, and covers an area 143 × 146 km. The scallops around volcanoes such as these are believed to be caused by several catastrophic landslides; over 80 such scalloped-margin volcanoes have been identified on Venus. The blocky aprons are debris that came down the steep slopes and out onto the plains, carried by their momentum. At the base of the east-facing or largest scallop on the volcano is what is interpreted to be a large block, 8–10 km in length. Resolution of the Magellan data is about 120 m.

One of the most surprising volcanic features discovered by Magellan were channels, a few kilometers wide and hundreds of kilometers long, that snake their way across the surface of Venus (Figure 4.10) (Baker *et al.*, 1992). These channels could not have been formed by water, which has not existed on the surface for

**Figure 4.10.** This Magellan SAR image shows a 200-km portion of a sinuous channel on Venus. The sinuous channel is about 2 km wide. Channels such as this are common on the plains of Venus, and are likely to have been formed by highly fluid lavas, perhaps of unusual composition. The image is centered at 49°S latitude, 273°E longitude, and covers an area of 130 × 190 km. Resolution of the Magellan data is about 120 m.

billions of years, if ever. They have formed by lava of some sort, a lava so fluid that it behaved like water. Volcanologists studying these channels have proposed a number of compositions that could be responsible for the channels, including carbonate rich or sulfur lavas and ultramafic silicate melts. Exotic compositions like these are necessary to explain how the lava could travel so far without cooling. Others have suggested that the channels were formed by erosion of the surface, similar to lunar rilles on the moon. One useful aspect of these channels, aside from their implications about compositions on Venus, are the fact that the channels extend for long distances, allowing them to be used as a time marker. The assumption is that the channel formed over a relatively short period of time. Therefore, geologic units that may not touch each other, but have clear age relationships with the channel, can then be related to each other. For example, the channel may superpose one lava flow, but be overlain by another. We could then interpret the latter flow to be older than the former, although they may not be in direct contact. In addition, some of the channels now trend uphill, indicating that the surface has deformed after they formed (we assume that lava, like most liquids, tends to flow downhill!).

The last type of unusual volcanic feature is coronae, large (>100 km across) circular features (Figure 4.11) (Pronin and Stofan, 1990). Over 400 coronae have been identified on Venus; the largest is 2,500 km across (Stofan et al., 2001b). These features are formed by rising blobs or a stream of hot material that bow up and deform the surface producing a ring of concentric ridges. The features tend to be raised at least 1 km above the surrounding plains, but also can be rimmed depressions or rimmed plateaus. Most coronae are located along rift systems, areas where the crust of the planet has pulled apart. The pulling apart of the crust probably makes it easier for hot material to come near the surface to form a corona. These features also form at hotspot rises, overlapping in time with the formation of large volcanoes. Coronae often have small shields in their interiors and intermediate volcanoes, and some are surrounded by very extensive lava flows (>200 km long). The cause of all the volcanism is melting produced by the thermal anomaly that forms the structures of the coronae.

## 4.2   THE VOLCANIC HISTORY OF VENUS

The volcanic landforms we can map on the surface of Venus provide clues to the overall geologic evolution of the planet. Volcanism is produced by heat originating in the interior of the planet; in large silicate planets such as the terrestrial planets, most of that heat is generated by the decay of radioactive elements (K, Th, U) in the interior. The larger a planet is, the more heat will be generated in the interior, and thus the more active the surface will be. Earth is obviously very geologically active, with about 25% of the internal heat generated driving the process of plate tectonics. Venus, so similar in size to the Earth, should be generating enough heat to drive the current geologic activity on the surface (Solomon and Head, 1982).

On most planets, we use the density of impact craters to determine the age of surface geologic units. The older a surface is, the more impact craters it will have

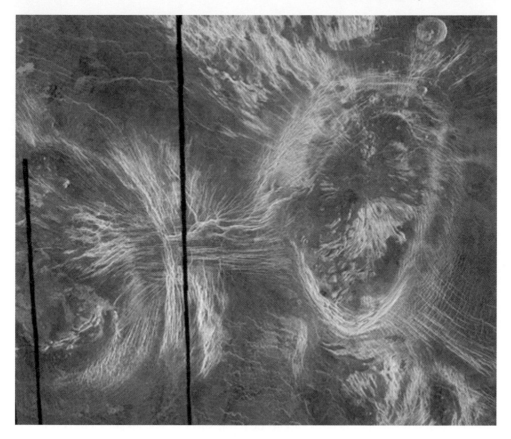

**Figure 4.11.** This Magellan SAR image mosaic shows two coronae, large circular or oval structures first identified in Soviet radar images of Venus. The image is centered at 49°N latitude, 2°E longitude. The feature on the left, Bahet Corona, is about 230 km long and 150 km across. Part of Onatah Corona (about 350 km in diameter) is to the right of Bahet. The coronae are surrounded by a concentric ring of ridges and troughs. The centers of the coronae have radial fractures, volcanic shields and domes, and flows. Coronae are interpreted to form due to the upwelling of hot material from the interior of Venus. Resolution of the Magellan data is about 120 m.

collected over time. Given information we have on the flux of impactors (comets, asteroids) through the solar system, and information on the age of surface rocks returned from the Apollo missions to the moon, we can assign absolute age dates to surface units on planetary bodies. On Venus, however, we have a problem. There are relatively few impact craters, approximately one per million square kilometers. In addition, the population of craters cannot be distinguished from a random population, making any attempt to use variations in impact crater density on geologic units statistically ambiguous (Schaber *et al.*, 1992). Overall, estimates of the crater retention age of the surface range from 300 million years to 1 billion years (the cratons or oldest continental areas on Earth have a similar crater retention age) (McKinnon *et al.*, 1997). Therefore, we do not appear to have any easy window into the early history of Venus (the first 3.5 billion years!). In addition, we cannot say how active the planet currently is, without future mission data. Techniques to measure the current level of activity on Venus would include a network of seismometers, which would be very difficult to operate for the months necessary to detect venusquakes given the high temperature and pressure on the surface of Venus.

Models to explain the statistically random population of impact craters range from a catastrophic episode of volcanism resurfacing the entire planet ("wiping the slate clean", with the impact population then slowly accumulating with little further geologic activity) (Strom *et al.*, 1994), to a more uniform in time resurfacing through smaller scale patch-style volcanism (Phillips *et al.*, 1992). A variation on the catastrophic model suggests that certain types of features tended to form at specific times in Venus' history, for example, all the large volcanoes postdate formation of the venusian plains (Basilevsky *et al.*, 1997). The alternative patch-resurfacing model would favor the plains deposits being interfingered with large volcano flows (Guest and Stofan, 1999). If there were catastrophic outpourings of lava, enough gases would have been released into the atmosphere to actually increase the surface temperature by over 400°C for a period of time (Bullock and Grinspoon, 2001). Detailed mapping of the surface is ongoing, and thus far has not produced any evidence that conclusively favors one model. It will take new data from future missions, including seismic data to reveal the internal layering of Venus and age dates from rocks returned to Earth from the surface of Venus, to resolve the scientific debate.

What we can say about the history of Venus, is that for about the last 700 million years there has been a great deal of surface volcanism. The range of landforms we see (such as ridged flows and domes) indicates that there are likely to have been more evolved compositions erupted, indicating remelting of crustal material. Many of the volcanic landforms, such as the large volcanoes and coronae, appear to have been long-lived, similar to hotspots on Earth, allowing time for evolution of magma chambers, and the likely eruption, again, of more evolved compositions of lava. There appears to have been a limited amount of explosive volcanism, as evidenced by some peculiar deposits at the summits of a few volcanoes (Campbell and Rogers, 1994). The water content of magmas is likely to be low, but water can be tied up in rocks and magma inside the planet even if it is too hot on the surface for liquid water. The combination of water with other gases such as $CO_2$ and $SO_2$ in erupting lavas must have been sufficient to allow some explosive activity, producing some of these peculiar deposits. These deposits may have been erupted very recently or may be millions of years old; we have no data on their age at this time.

The only evidence we have to possibly support active volcanism on Venus are measurements of declining $SO_2$ in the atmosphere, first measured in 1978, which some scientists interpreted to be consistent with a recent eruption(s) that ceased sometime prior to 1978 (Esposito, 1984). Over the few years that Magellan was in orbit (1990–1994), no changes that could be interpreted as an eruption could be seen in the radar images of flow fields around volcanoes. However, there could have been small eruptions that produced changes below the resolution of the radar ($\sim$120 m). In addition, if a lava flow was erupted onto a surface with similar radar characteristics (i.e., degree of blockiness), it is unlikely to have been detected with the Magellan radar. On Earth, there are generally more than 10 volcanoes erupting at any given time. If Venus is even half as active as Earth, perhaps a future orbiting spacecraft could detect atmospheric changes from an erupting volcano.

How does the study of Venusian volcanoes help us to understand how volcanoes work? It turns out that some of the predictions made prior to Magellan on what the characteristics of volcanoes on Venus should be, based on what we know about volcanoes on Earth, were proven to be wrong. Lava flows were much longer than predicted, and we still have no conclusive theory for how unusual features such as the steep-sided domes and channels have formed, and we have been very surprised at the extremely large number of small shields (Saunders *et al.*, 1991; Saunders *et al.*, 1992). These new data make us go back to the drawing board on some of the most basic models, which are important to us here on Earth. For example, if you live near a large volcano such as Mount Vesuvius here on Earth, you want to know how far the flows might go if the volcano erupts. The longer flows on Venus tell us that we do not yet have a complete understanding of the formation of lava flows on planetary surfaces. In addition, we know that emissions from erupting volcanoes, not to mention from human activity, have the potential ability to change our climate. Understanding the effects of volcanism on the Venusian atmosphere—such as the possible large temperature increase from global resurfacing discussed above—are helping us to explore the stability of terrestrial planet atmospheres. We will use the Magellan data of volcanic features on Venus to refine models based on our terrestrial experience.

## 4.3   FUTURE EXPLORATION OF VENUS

We have many unanswered questions about Venus that have resulted from study of the rich Magellan data set. Many of the questions center on the geologic evolution of the surface and interior, and why the planet has evolved so differently from Earth. Answering these questions will be challenging, due to the unusual and harsh conditions that exist on the surface of the planet. In the short-term, missions to Venus can address some fundamental questions about the composition of the atmosphere, which gives insights into how the planet has evolved. We also have the technology to send missions to the surface, as long as they don't have to last much longer than an hour. We only have the technology to insulate a spacecraft against the extreme heat of the surface of Venus for a few hours. This type of mission could provide visible descent imaging of the surface (we should be able to see the surface of Venus with conventional imaging cameras at about 10–20 km above the surface), information on surface composition, and data on the composition of the atmosphere near the surface. The near-surface visible wavelength descent images would help us tie the characteristics of the landing site (images and surface compositional information) to the Magellan data, providing "ground truth" for better interpretation of the radar images. The near surface atmosphere, which reacts chemically with the surface, has not been measured by any previous probes. However, to really get at why Venus has evolved so differently than Earth, we need missions to Venus that involve advances in technology.

Seismic data, measurements of the shaking of the planet due to an earthquake, are critical measurements that have helped us to understand much about the Earth. Seismic data provides a detailed picture of the internal layering within a planet, information on the solid versus the liquid state of the layers, and even information on the internal composition of the planet. Very large quakes (>magnitude 7) set up a ringing of the planet that allows the very deepest structure to be measured. Much of our debate on Venus has come to center on models that involve the interior structure being very different from that of the Earth, and models have been developed that make very specific predictions on the internal layering of Venus. Therefore, seismic data for Venus would allow us to differentiate between ideas of whether the planet has a history characterized by catastrophic overturn, whether the interior has much thicker layers than that of the Earth, and whether the core is very different from the Earth's (Venus lacks a detectable magnetic field, and our theory of magnetic field formation suggests that Venus should have a solid core). However, the problem is that to make these detailed seismic measurements, a network of at least three seismometers placed fairly far apart is required, along with an orbiter to communicate with them. In addition, you need them to last for at least a year, due to the fact that large magnitude quakes are probably not as frequent as on Earth. The electronics required to run the seismometers and to send data from the surface to an orbiter would have to either be kept cool by a radioactively powered cooler, or electronic components that could operate at the extreme Venus surface temperatures would need to be developed. All of this is technologically difficult and likely to be expensive! However, the data set would revolutionize our view of the evolution of Earth-like planets.

Another critical data set is the information about the composition of the surface, and how much it varies. On planets without cloud-obscured surfaces such as Mars, we can send orbiting instruments that map the composition of the surface on a global scale. However, the thick cloud cover on Venus prevents such orbital measurements from being made. Therefore, we must rely on missions to the surface to measure the composition of different terrain types, in order to understand the range of compositions that exist and what they imply for the overall history of the planet. The Venera landers returned bulk element compositions of rocks in the lowland plains that are similar to terrestrial basalts. In the future, we would like to measure both what elements are contained in Venusian surface rocks (such as isotopes of silicon and aluminum that tell you about the planet's accretion) to the mineralogy of the rock (e.g., minerals that crystallize out of a magma can tell you about what volatiles were in the magma). We would use similar instruments to those sent to the surface of Mars (using techniques such as x-ray diffraction, x-ray fluorescence, spectrometry; see Chapter 7), but they would have to be altered to operate under Venusian surface conditions. High priority targets for future lander missions include tessera terrain, highly deformed regions that might be composed of older, more continent-like materials, sites of evolved volcanic composition, and sites of hotspot volcanism. The problem with targeting sites such as channels or unusual lava flows is that they are relatively small (<10s of kilometers). It is very difficult to target a lander to a small region on a planet's surface; landers generally have landing ellipses of 100s of kilometers.

In order to really understand how Venus compares with the other terrestrial planets, we will need to bring a sample from the surface of Venus back to Earth. On Earth, we have laboratories with very specialized instruments, which can measure to very high precision isotopic compositions of rocks or soil. Different isotopes can be used to help to constrain models of the formation of all of the terrestrial planets, to tell us how both the interior and surface of the planet have evolved, and the precise age of the material returned. We already have samples of the Moon and Mars (the Mars meteorites) to compare to terrestrial rocks; venusian rocks would fill an important piece of the puzzle (along with returned samples from Mercury someday). Due to the large size of Venus, a sample return mission is technologically challenging! It is easy to get down to the surface, but hard to launch back into space from the surface of a planet with gravity similar to that on Earth. This mission truly holds the key to solving many of the puzzles of Venus, but will undoubtedly be decades in the future.

It is frustrating to think that answers to the questions that excite planetary scientists the most are so difficult to obtain. However, studying Venus will come eventually, because it does hold such key information to understanding why Earth is so unique in our solar system. When we look out into space, we wonder how many Earth-like planets there are around other stars, planets with conditions such that water, so key to life on this planet, is stable on the surface. We need to understand why the conditions for this were only "just right" here on Earth. Venus, our 'twin', holds many of the answers still, beneath its dense and toxic clouds.

## 4.4  REFERENCES

Cattermole, P. (1994) *Venus: The geological story*. UCL Press, London, 250 pp.

Baker, V.R., Komatsu, G., Parker, T.J., Gulick, V.C., Kargel, J.S., and Lewis, J.S. (1992) Channles and valleys on Venus: Preliminary analysis of Magellan data. *J. Geophys. Res.*, **97**, 13421–13444.

Barsukov, V.L., Basilevsky, A.T., Kuzmin, R.O., Pronin, A.A, Kryuchkov, V.P., Nikolaeva, O.V., Chernaya, I.M., Burba, G.A., Bobina, N.N., Shashkina, V.P. *et al.* (1984) Preliminary evidence on the geology of Venus from radar measurements by the Venera 15 and 16 probes. *Geokhimia*, **12**, 1811–1820. (*Geochem. Int. Engl. Transl.*, **22**, 135–143, 1985).

Basilevsky, A.T., Head, J.W., Schaber, G.G., and Strom, R.G. (1997) The resurfacing history of Venus. In: S.W. Brougher, D.M. Hunten, and R.J. Phillips (eds), *Venus II*. University of Arizona Press, Tucson, AZ, pp. 1047–1085.

Basilevsky, A.T., Pronin, A.A., Ronca, L.B., Kryuchkov, V.P., Sukhanov, A.L., and Markov, M.S. (1986) Styles of tectonic deformation on Venus: Analysis of Veneras 15 and 16 data. *J. Geophys. Res.*, **91**, 399–411.

Bridges, N.T. (1997) Ambient effects on basalt and rhyolite lavas under Venusian, subaerial and subaqueous conditions. *J. Geophys. Res.*, **102**, 9243–9255.

Bullock, M.A. and Grinspoon, D. (2001) The recent evolution of climate on Venus. *Icarus*, **150**, 19037–19048.

Campbell, B.A. and Rogers, P.G. (1994) Bell Regio, Venus: Integration of remote sensing data and terrestrial analogs for geologic analysis. *J. Geophys. Res.*, **99**, 21153–21171.

Crumpler, L.S., Aubele, J.C., Senske, D.A., Keddie, S.T., Magee, K.P., and Head, J.W. (1997) Volcanoes and centers of volcanism on Venus. In: S.W. Brougher, D.M. Hunten, and R.J. Phillips (eds), *Venus II*. University of Arizona Press, Tucson, AZ, pp. 697–756.

Esposito, L.W. (1984) Sulfur dioxide shows evidence of active Venus volcanism. *Science*, **223**, 1072–1074.

Florensky, K.P., Basilevsky, A.T., Burba, G.A., Nikolaeva, O.V., Pronin, A.A, Selivanov, A.S., Naraeva, M.K., Panfilov, A.S., and Chemodanov, V.P. (1983), Panormas of Vener 9 and 10 Landing sites. In: D.M. Hunten, L. Colin, T.M. Donahue, and V.I. Moroz (eds), *Venus*. University of Arizona Press, Tucson, AZ, pp. 137–153.

Grinspoon, D.H. (1998) *Venus Revealed*. Perseus Publishing, New York, 355 pp.

Guest, J.E. and Stofan, E.R. (1999) A new view of the stratigraphic history of Venus. *Icarus*, **139**, 55–66.

Guest, J.E., Bulmer, M.H., Aubele, J., Beratan, K., Greeley, R., Head, J.W., Michaels, G., Weitz, C., and Wiles, C. (1992) Small volcanic edifices and volcanism in the plains of Venus. *J. Geophys. Res.*, **97**, 15949–15966.

Head, J.W., Crumpler, L.S., Aubele, J.C., Guest, J.E., and Saunders R.S. (1992) Venus Volcanism: Classification of Volcanic Features and Structures, Associations, and Global Distribution from Magellan Data. *J. Geophys. Res.*, **97**, 13153 13197.

Lancaster, MG., Guest, J.E., and Magee, K.P. (1995) Great lava flow fields on Venus. *Icarus*, **118**, 69–74.

Masursky, H., Kaula, W.M., McGill, G.E., Pettengill, G.H., Schaber, G.G., and Schubert, G., (1980) Pioneer Venus radar results: Geology from images and altimetry. *J. Geophys. Res.*, **85**, 8232–8260.

McKinnon, W.B., Zahnle, K.J., Ivanov, B.A., and Melosh, H.J. (1997) Cratering on Venus: Models and observations. In: S.W. Brougher, D.M. Hunten, and R.J. Phillips (eds), *Venus II*. University of Arizona Press, Tucson, AZ, pp. 969–1014.

Pavri, B., Head, J.W., Klose, K.B., and Wilson, L. (1992) Steep-sided domes on Venus: Characteristics, geologic setting, and eruption conditions from Magellan data. *J. Geophys. Res.*, **97**, 13445–13478.

Pettengill, G.H., Eliason, E., Ford, P.G., Loriot, G.B., Masursky, H., and McGill, G.E. (1980) Pioneer Venus radar results: Altimetry and surface properties. *J. Geophys. Res.*, **85**, 8261–8270.

Phillips, R.J., Raubertas, R.F., Arvidson, R.E., Sarkar, I.C., Herrick, R.R., Izenberg, N., and Grimm, R.E. (1992) Impact craters and Venus resurfacing history. *J. Geophys. Res.*, **97**, 15923–15948.

Pronin, A.A. and Stofan, E.R. (1990) Coronae on Venus: Morphology, classification and distribution. *Icarus*, **87**, 452–474.

Saunders, R.S., Spear, A.J., Allin, P.C., Austin, R.S., Berman, A.L., Chandlee, R.C., Clark, J., Decharov, A.V., De Jong, E.M., Griffith, D.G., *et al.* (1992) Magellan mission summary. *J. Geophys. Res.*, **97**, 13067–13090.

Saunders, R.S., Arvidson, R., Head, J.W., Schaber, G., Solomon, S., and Stofan, E. (1991) Magellan: A first overview of Venus geology. *Science*, **252**, 249–252.

Schaber, G.G., Strom, R.G., Moore, H.J., Soderblom, L.A., Kirk, R.L., Chadwick, D.J., Dawson, D.D., Gaddis, L.R., Boyce, J.M., and Russell, J. (1992) Geology and distribution of impact craters on Venus: What are they telling us? *J. Geophys. Res.*, **97**, 13257–13302.

Solomon, S.C., and Head, J.W. (1982) Mechanisms for lithospheric heat transport on Venus: Implications for tectonic style and volcanism. *J. Geophys. Res.*, **87**, 9236–9246.

Stofan, E.R., Smrekar, S.E., Bindschadler, D.L., and Senske, D.A. (1995) Large topographic rises on Venus: Implications for mantle upwelling. *J. Geophys. Res.*, **100**, 23317–23327.

Stofan, E.R., Anderson, S.W., Crown, D.A., and Plaut, J.J. (2000) Emplacement and composition of steep-sided domes on Venus. *J. Geophys. Res.*, **105**, 26757–26772.

Stofan, E.R., Guest, J.E., and Copp, D.L. (2001a) Development of large volcanoes on Venus: Constraints from Sif, Gula and Kunapipi Montes. *Icarus* **152**, 75–95.

Stofan, E.R., Tapper, S.W., Guest, J.E., Grindrod, P., and Smrekar, S.E. (2001b) Preliminary analysis of an expanded corona database for Venus. *Geophys. Res. Lett.*, **28**, 4267–4270.

Strom, R.G., Schaber, G.G., and Dawson, D.D. (1994) The global resurfacing of Venus. *J. Geophys. Res.*, **99**, 10899–10926.

Surkov, Yu.A. (1983) Studies of rocks by Veneras 8, 9, and 10. In: D.M. Hunten, L. Colin, T.M. Donahue, and V.I. Moroz (eds), *Venus*. University of Arizona Press, Tucson, AZ, pp. 154–158 (and other papers therein).

# 5

## The face of the Moon: Lunar volcanoes and volcanic deposits

*Lisa Gaddis* (US Geological Survey)

*The Moon has been mine as long as I can remember. Growing up in New Orleans, Louisiana, with its thick and sultry atmosphere, I could see the Moon clearly but I could not see many of the stars that populate the vivid night sky. I watched the Moon and I came to recognize its changes. Wherever I was, the Moon was always with me I could make wishes, I could imagine other worlds, and I could fall in love with the idea that we are not alone on Earth in space. When I was 11 years old, I watched a man step onto the Moon and I knew that I wanted to understand that place.*

*When I left New Orleans to go to Vassar College, I could not believe the brilliance of the night sky in Poughkeepsie, New York. Although I had already determined that I would study geology, I quickly decided that I would also learn as much as I could about the Moon and other planets. And I have. The Moon, especially its volcanic deposits, has been a major emphasis of my research since then. I have learned to understand how the colors of lunar materials behave at different wavelengths, how to interpret the composition of lunar materials, and how to relate that to the geologic history of the Moon. My tools are the computer, color digital images of the Moon, and mathematical methods for reducing those images to pictures that can be easily understood. I have worked with many of the world's premier planetary scientists, and through them I learned how and why planetary science is done. I have learned to recognize and nurture ability in many forms, and to build on my strengths as a communicator and a collaborator. I am now part of international discussions to plan a return to the Moon. I fully expect that my children and yours, or perhaps their descendents, will someday live and work on our Moon.*

## 5.1    THE MOON FROM EARTH: A SHORT HISTORY OF LUNAR OBSERVATION AND EXPLORATION

The Moon has been Earth's companion in space for a very long time, nearly 5 billion years, and it has looked like it does today for much of that time (Figure 5.1). Ever since there have been human eyes on Earth to look up at the Moon, people have been marveling at it and wondering where it came from and what it means. People have noted that we always see the same face or side of the Moon, that the Moon changes its appearance daily, and that it does this several times a year. These observations of the Moon have influenced human folklore and culture, fertility cycles, planting and growing schedules, timing of battles and wars, and the development of our measurement of time, from our annual calendars down to our daily schedules.

For many years, we were limited to the use of our eyes to study the Moon. In 1609, the Italian Galileo Galilei popularized the use of the telescope. Between then and the present day the telescope, in its various forms, has been used to study, map, and name the features of the Moon. In addition to visible wavelengths, both shorter and longer wavelengths (from ultraviolet to infrared to radar, or 300-nm to 70-cm wavelengths) have been used to study the Moon. We can see that the Moon has vast regions of rough, bright deposits, called "terrae" or highlands, and smooth, dark, lower-elevation regions, called "mare" or "maria" (plural) because they resembled oceans. Several spacecraft with cameras and telescopic lenses were sent to study the Moon prior to the Apollo landings, including the Soviet Zond (1965–1970) and Luna (1959–1972) missions, and the US Ranger (1962–1965), Surveyor (1966–1968), and Lunar Orbiter (1966–1968) missions (Table 5.1; see also Heiken et al., 1991; Spudis, 1996). From those photographs and remotely acquired data on lunar surface physical properties we learned that the surface of the Moon was not covered with thick layers of dust that we might sink into upon landing. We could identify smooth, boulder-free locations where it might be possible to land safely. With the beautiful photos obtained by the Lunar Orbiter missions, we were able to see details of the entire surface of the Moon, and thus to create geologic maps that are still in use today. These early robotic missions to the Moon gave us an understanding of the broad geologic history of the Moon, and the knowledge of how to get there safely and what to expect when we returned in the future. These data also caused us to ask even more questions about how the Moon came to be as we see it now.

As part of the famous lunar walks by astronauts on the Moon in the late 1960s and early 1970s, the Apollo missions returned ~380 kg of rocks and soils. The robotic Soviet Luna missions in the 1970s returned an additional ~276 kg of lunar soil. These samples told us much about the Moon, especially bringing home the fact that it has had an active, if ancient, geologic history. Rocks from the Moon allowed us to understand that much of it is very old, up to 4.3 billion years old and that it has been very dry. The internal processes of melting and chemical differentiation have played an important role in modifying the heavily cratered lunar surface (Neal and Taylor, 1992; Shearer and Papike, 1993). Four major classes of lunar materials were

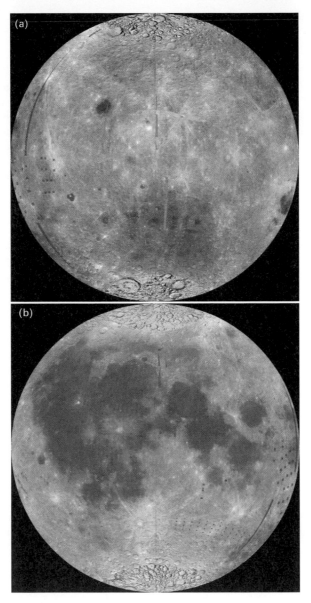

**Figure 5.1.** The full Moon as viewed by Clementine in 1994. These views are mosaics made up of thousands of small Clementine "postage stamp" images obtained at a near-infrared wavelength of 750 nm and processed to 100 m/pixel. The dark areas are covered with lava flows or maria, and the rough, brighter areas are lunar highlands. Note that most of the maria are seen on the lunar near side. The Clementine albedo data are superimposed on a shaded relief view of the Moon for clarity. Vertical stripes in the Clementine image are areas of missing data in the mosaic. (a) Far side, Lambert Azimuthal Equal Area projection, centered at 180° longitude. (b) Near side, same projection, centered at 0° longitude.
Courtesy of Matthew Staid, USGS.

**Table 5.1.** Lunar exploration missions.

| Mission | Launch Date | Country | Type |
|---|---|---|---|
| Luna 1 | January 1959 | USSR | Fly-by |
| Luna 2 | September 1959 | USSR | Hard lander* |
| Luna 3 | October 1959 | USSR | Fly-by |
| Ranger 3 | January 1962 | USA | Hard lander |
| Ranger 4 | April 1962 | USA | Hard lander |
| Ranger 5 | October 1962 | USA | Hard lander |
| Luna 4 | April 1963 | USSR | Fly-by |
| Ranger 6 | January 1964 | USA | Hard lander |
| Ranger 7 | July 1964 | USA | Hard lander |
| Ranger 8 | February 1965 | USA | Hard lander |
| Ranger 9 | March 1965 | USA | Hard lander |
| Luna 5 | May 1965 | USSR | Soft lander** |
| Luna 6 | June 1965 | USSR | Fly-by |
| Zond 3 | July 1965 | USSR | Fly-by |
| Luna 7 | October 1965 | USSR | Soft lander |
| Luna 8 | December 1965 | USSR | Soft lander |
| Luna 9 | January 1966 | USSR | First soft landing |
| Luna 10 | March 1966 | USSR | First lunar orbiter |
| Surveyor 1 | May 1966 | USA | Soft lander |
| Lunar Orbiter 1 | August 1966 | USA | Orbiter |
| Luna 11 | August 1966 | USSR | Orbiter |
| Surveyor 2 | September 1966 | USA | Lander |
| Luna 12 | October 1966 | USSR | Orbiter |
| Lunar Orbiter 2 | November 1966 | USA | Orbiter |
| Luna 13 | December 1966 | USSR | Soft lander |
| Lunar Orbiter 3 | February 1967 | USA | Orbiter |
| Surveyor 3 | April 1967 | USA | Soft lander |
| Lunar Orbiter 4 | May 1967 | USA | Orbiter |
| Surveyor 4 | July 1967 | USA | Lander |
| Explorer 35 | July 1967 | USA | Orbiter |
| Lunar Orbiter 5 | August 1967 | USA | Orbiter |
| Surveyor 5 | September 1967 | USA | Soft lander |
| Surveyor 6 | November 1967 | USA | Soft lander |
| Surveyor 7 | January 1968 | USA | Soft lander |
| Luna 14 | April 1968 | USSR | Orbiter |
| Zond 5 | September 1968 | USSR | Fly-by, return |
| Zond 6 | November 1968 | USSR | Fly-by, return |
| Apollo 8 | December 1968 | USA | Human orbiter, return |
| Apollo 10 | May 1969 | USA | Human orbiter, return |
| Luna 15 | July 1969 | USSR | Sample return |
| Apollo 11 | July 1969 | USA | Human lander, return |
| Zond 7 | August 1969 | USSR | Fly-by and return |
| Apollo 12 | November 1969 | USA | Human lander, return |
| Apollo 13 | April 1970 | USA | Human lander (aborted) |
| Luna 16 | September 1970 | USSR | First sample return |

| Mission | Launch Date | Country | Type |
|---------|-------------|---------|------|
| Zond 8 | October 1970 | USSR | Fly-by and return |
| Luna 17 | November 1970 | USSR | Rover |
| Apollo 14 | January 1971 | USA | Human lander, return |
| Apollo 15 | July 1971 | USA | Human lander, return |
| Luna 18 | September 1971 | USSR | Sample return |
| Luna 19 | September 1971 | USSR | Orbiter |
| Luna 20 | February 1972 | USSR | Sample return |
| Apollo 16 | April 1972 | USA | Human lander, return |
| Apollo 17 | December 1972 | USA | Human lander, return |
| Luna 21 | January 1973 | USSR | Rover |
| Luna 22 | May 1974 | USSR | Orbiter |
| Luna 23 | October 1974 | USSR | Sample return |
| Luna 24 | August 1976 | USSR | Sample return |
| Muses A | January 1990 | Japan | Orbiter |
| Galileo | December 1990 | USA | Fly-by |
| Galileo | December 1992 | USA | Fly-by |
| Clementine | January 1994 | USA | Orbiter |
| Lunar Prospector | January 1998 | USA | Orbiter |

\* Hard landers, intended to crash onto the lunar surface.
\*\*Soft landers, intended to be guided onto the lunar surface gently.

identified in these samples, including ancient highlands rocks from the lunar crust, basaltic volcanic rocks (including lava flows and pyroclastic or "lava-fountain" deposits), broken and melted rocks from the numerous impacts that have marked the lunar surface over the eons, and the "regolith" or pulverized, fine lunar soil. The character and origin of lunar volcanoes and their volcanic deposits are the subjects of this chapter.

Exploration of the Moon has taken a great leap forward in the last decade. Several spacecraft, including Galileo (1990, 1992), Clementine (1994), and Lunar Prospector (1998) orbited the Moon and returned images and data that allow us to understand better the elevations of lunar features, the internal behavior of the Moon, the chemistry and mineralogy of lunar rocks and soils, the presence of a weak magnetic field and an inferred metallic core, and the number and timing of impact events in our portion of the solar system. New missions are also underway or are being planned, including the SMART-1 (launched by the European Space Agency on 27 September, 2003), Lunar-A (Japan, 2004), and Selene (Japan, 2005). We have learned much about the Moon in 40 years, but we still have much more to learn. This chapter will first present an overview of what we know about lunar volcanism and its history, will describe some of the basic differences between volcanoes and volcanic rocks on the Earth and Moon, and will end by describing what we don't know about the Moon and where our quest for answers may take us.

## 5.2   THE FACE OF THE MOON

Looking at the Moon, one can hardly miss the countless circular features called craters. Although we could see these craters clearly from Earth and space, and we could recognize and study similar features on the Earth, the origin of these large round holes was the subject of heated debate for more than 100 years. Although we now know that they were formed by impacts, a volcanic origin for craters on the Earth and moon was vehemently defended by many well-respected scientists as recently as 30 years ago. The fact that many of the large craters on the moon were filled with dark mare deposits merely fueled this controversy. Indeed, when the idea that impact processes had formed the lunar craters began to take hold among planetary scientists, it was thought by some that these holes were filled with "impact melt" (fluidized rock melted by the energy of the impact) instead of by lava. This reasoning has been extended to the heavily cratered planet Mercury (Robinson and Lucey, 1997), where dark, smooth deposits inside craters are sometimes believed to have formed by impact melting rather than by volcanism. Although impact melts are now recognized on the Moon, they are much less voluminous and widespread than volcanic deposits. We now know that the vast, dark, smooth surfaces on the Moon are ancient volcanic deposits. Although volcanic deposits occupy only about 16% of the lunar surface area, they have played a major role in the geologic history of the Moon.

Even before we visited the Moon and returned samples, we could see strong evidence for the presence of lava flows. The dark, smooth deposits or maria that seemed to fill huge basins and craters were apparently very fluid at the time of their emplacement. Using telescopes, we could see occasional raised edges marking the margins of huge lunar lava flows. We understood that we were most likely seeing multiple layers of lava flows in the filled impact basins and craters (Figure 5.2). We learned to count impact craters large and small as a measure of the relative ages of these deposits, and we could see that there were subtle differences in darkness (or albedo) among lunar maria. Indeed, the simple method of counting craters to date planetary surfaces was developed for geologic studies of the Moon and has proved to be one of the most important tools of the planetary geologist for studying the terrestrial planets (Wilhelms, 1997; Hiesinger *et al.*, 2003). One measures a unit surface area and then the diameters of each primary impact crater within that unit. Based on a model of the population of lunar impact craters over time (the "lunar production function"), these data are related to the time a unit has been exposed on the lunar surface. These relative ages must be calibrated by linking ages of lunar samples to crater counts. Even so, resulting errors in relative age of 10% to 25% are estimated for most craters. After we examined the lunar samples we began to assign real dates to some of the lunar maria, and we realized that differences in darkness were related to compositional differences among maria, particularly in the amounts of elements such as iron and titanium. Mare deposit compositions also indicated to us that lunar lava flows were very runny and that they flowed somewhat like shampoo, probably spreading relatively quickly and quietly across large areas of the lunar surface. More recently, infrared color data of the Moon returned from

**Figure 5.2.** Apollo 15 fly-by (oblique) view of lava flows in Mare Imbrium (25°N, 330°E), shown extending from lower left to upper right. These flows are superimposed on a cratered, wrinkled mare plains unit and are probably about 35–50 m thick, extend up to 500 km long, and are thought to be more than 2.5 billion years old. Looking at the lower left of the image, note that the flow likely traveled along and then down a scarp or ridge in the lava plains. Near the center of this image, the flow is cut by a prominent ridge and thus probably formed before the ridge. The dark area at the right edge of the image is part of the Apollo 15 spacecraft. Apollo 15 metric camera photo, AS15-M-1556.

Galileo and Clementine showed striking differences among the lunar maria and made us realize how few volcanic deposits had been sampled and how much more diverse lunar volcanism has been than we previously understood (Gaddis *et al.*, 1985; Staid and Pieters, 2001; Gaddis *et al.*, 2003).

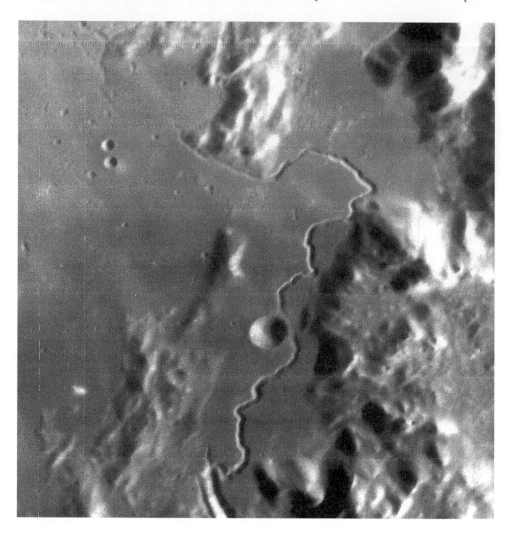

**Figure 5.3.** A view of Hadley Rille (25°N, 3°E) and the Apollo 15 landing site (near the "elbow" of the rille upper right of the image) on the south-eastern edge of Mare Imbrium as observed by the orbiting Lunar Orbiter IV camera. At more than 120 km long, 1500 m across, and 300 m deep, Hadley Rille begins in the bottom left of the image and extends north-eastward, curving abruptly around to the west, and then disappearing from sight into the maria of Palus Putredinis in the upper left of the image. Lava channels in Hawaii are much smaller, usually less than 10 km long and up to 100 m wide at most.
From Lunar Orbiter IV-110-H1.

Sinuous rilles are another type of volcanic feature found in lunar maria. They often originate in irregular fractures or craters in the highlands or maria, and they always follow the terrain downward into the maria. Hadley Rille in south-eastern Mare Imbrium was visited by Apollo 15 astronauts (Figure 5.3). One of the largest of the lunar rilles, Hadley is more than 100 km long, up to 3 km wide, and 1 km deep. (For comparison, the Grand Canyon, Arizona, is about 450 km long, up to 29 km wide, and 1.8 km deep.) At first glance, these relatively narrow, winding channels on the Moon resemble water-cut river or stream channels on the Earth, but they are now known to have been carved by hot, turbulently flowing lava. In some areas, lunar sinuous rilles are associated with round, crater-like features that probably mark the locations of collapsed lava tubes much like the features seen in the volcanic Hawaiian islands. Lunar sinuous rilles confirm our understanding of the lunar maria as having been formed by the eruption of large volumes of very fluid lava.

Although the maria and lava flows are by far the most common form of lunar volcanic deposits, an intriguing variety of other volcanic features are also seen on the Moon (Head and Wilson, 1992). Among the most interesting are those found near the relatively few observable vents of lunar volcanic eruptions. Although much lower and often broader than similar features on Earth, we do indeed recognize lunar domes, cones, and blankets. Whereas the maria were probably erupted relatively quietly as massive flows from long, narrow fractures in the lunar crust, the domes, cinder cones, and ash blankets probably formed by low-volume eruptions from small, sometimes oddly shaped holes or cracks. Lunar domes and cones, such as those in the Marius Hills region of the Moon (Figure 5.4), are steeper than basalt shield volcanoes on Earth and may have formed by quick, short eruptions, perhaps with alternating layers of lava and ash. These features are commonly seen in clusters or fields of volcanoes within the lunar maria, and they are likely to mark ancient volcanic vents that were largely covered by younger lava flows.

Volcanic blankets or "mantles" on the Moon were formed by lava-fountain or pyroclastic eruptions in which small amounts of volcanic material mixed with gases spew from a single vent. Both small and large lunar pyroclastic deposits are known. The small deposits are often around holes aligned along cracks in crater floors (Figure 5.5), and they are thought to have formed from the periodic eruption of fragmented basalt and possibly ash, much like Vulcanian eruptions on Earth (Head and Wilson, 1979). Many lunar pyroclastic deposits are much larger and tend to be observed in the highlands along the margins of major lunar maria (Figure 5.6). The large pyroclastic deposits probably formed through a lava-fountain process, with a more continuous flow of magma over a longer period of time, similar to Strombolian-style eruptions on Earth. The numerous round glass beads in lunar samples were formed during large lunar pyroclastic eruptions and widely dispersed across the Moon (Adams *et al.*, 1974). About 20 varieties of these glasses are known and, in shades of purple, red, orange, yellow, green, and black, they are among the most colorful of lunar specimens (Delano, 1986). It is likely that lunar volcanic glasses formed as tiny particles of liquid lava were thrown into the cold lunar atmosphere and frozen there, falling to the surface as round, clear glass beads and droplets. The composition of these tiny beads is very primitive, or similar to that of the oldest, most basic of lunar materials, so they help us to understand what the deep interior of the Moon was like when lunar volcanism began (Shearer and Papike, 1993).

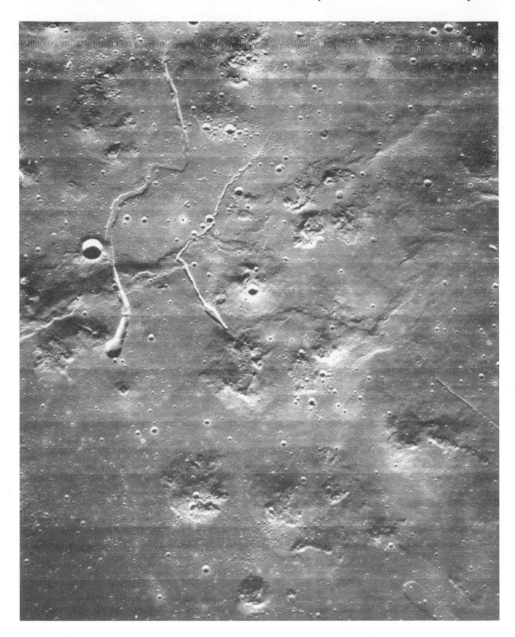

**Figure 5.4.** A Lunar Orbiter V view of a portion of the Marius Hills region of the Moon (14°N, 304°E), showing numerous volcanic cones and domes, two sinuous rilles, and a ridge. More than 250 domes, 60 cones, and 20 sinuous rilles are observed in this region. The domes are typically less than 300 m high and 3 km in diameter. The round "cobra-head" feature associated with the sinuous rille at the center left of the image is about 2.5 km in diameter. The image is 80 km across and north is toward the upper right.
Lunar Orbiter V frame 215-M.

**Figure 5.5.** Alphonsus crater floor (diameter 118 km; 13.7°S, 357°W) with at least 11 small pyroclastic deposits, often called "dark halo craters", aligned along linear fractures. The central peak of Alphonsus, probably formed by rebound and uplift shortly after the crater formed, is 9 km in diameter and 1.1 km high. The smooth, dark pyroclastic material probably consists of ash and fragments of local rock (possibly basalt).
Lunar Orbiter IV frame 108-H2.

## 5.3  LUNAR VOLCANIC ERUPTIONS

The observed characteristics of the major lunar volcanic deposits, especially the fact that they are often smooth, dark, and expansive, lead us to believe that many of the lava flows were formed by high-volume eruptions of relatively long duration, much like flood-basalts on Earth (Head and Wilson, 1992). From sample studies and our relative age dating using crater counts, we understand that lunar lava flows did not simply pour out in one huge event. The volcanic evolution of the Moon occurred over a very long period of time, starting early in lunar geologic history. The long epoch of lunar volcanism was punctuated by periods of interior melting and ascent

**Figure 5.6.** The very large pyroclastic deposit at Rima Bode (10°N, 356°W), shown covering or "mantling" the rougher highlands in the central lunar near side. These dark pyroclastic materials are among the darkest materials seen on the Moon, and they probably contain a large amount of black glass beads. The larger crater at the lower right of the image, Bode A, is about 12 km in diameter.
Lunar Orbiter IV frame 109-H2.

of magma through cracks in the crust, eruption of magma and ash at the surface, and the emplacement of lavas with different compositions in different places at different times.

Studies of lunar volcanic glass beads tell us that they were formed in fountains of lava on the Moon. But what caused these eruptions to be so violent, especially when the vast majority of lunar volcanic eruptions must have been relatively peaceful? We know that the Moon is very dry compared to the Earth, where water and magma interact to create spectacular lava-fountain displays in places like Kilauea volcano on the island of Hawaii. Traces of gases on the surfaces of the lunar glass beads give us a clue to the answer to this question. We find sulfur and carbon monoxide on their surfaces, and this tells us that these were major components of the gases that propelled these materials into the very tenuous lunar atmosphere (consisting largely of He, Ar, Na, and K and weighing only $10^7$ g). Thus there must have been scattered small pockets of such gases deep within the Moon. Although small in volume, and perhaps relatively infrequently involved in lunar volcanic eruptions, these gases play a critical role in our understanding of lunar volcanic history and thermal evolution because they brought up material from deep within the Moon.

A striking aspect of the distribution of lunar maria is that they occur largely on the near side of the Moon, and thus they are readily visible to us from Earth. We had tantalizing views of the far side of the Moon previously, but until we received data from Clementine we did not fully appreciate just how rare volcanic eruptions were there. In addition to the spectacular color views of the lunar surface, Clementine returned gravity and topographic data that allowed us to understand the thickness of the lunar crust for the first time. It turns out that the lunar crust is thicker on the far side (average 70 km there) of the Moon than on the near side (average 55 km). On the far side of the Moon, rising lunar magma was much less likely to have penetrated through the thicker crust. Thus differences in lunar crustal thickness appear to have played a major role in the eruption of lunar lavas at the surface.

## 5.4   A SHORT STORY ABOUT A LONG TIME

Volcanism on the Moon has probably occurred since the earliest days of lunar crustal formation, perhaps as long as 4.3 billion years ago. Traces of iron enrichment in the lunar highlands suggest that volcanic materials were present during the earliest periods of bombardment of the surface of the Moon. The major phase of lunar mare volcanism appear to have occurred after this bombardment began to decline, perhaps 3.8 billion years ago, when deposits that were created were no longer fragmented and destroyed by impact cratering. Many of the oldest lunar maria had high concentrations of titanium, and they were erupted to form Mare Tranquillitatis and Mare Serenitatis from about 3.8–3.6 billion years ago. Following that time, from 3.6 to about 3.1 billion years ago, voluminous lavas with lower titanium contents were erupted in Mare Serenitatis, Mare Imbrium, and Oceanus Procellarum. This period represented the peak of lunar volcanism, which began to decline in volume and frequency until it stopped about 1 billion years ago. The youngest lunar maria, with crater-count ages of ~1.2 billion years, are observed in Oceanus Procellarum and there we also see the greatest diversity of compositions and eruptive episodes (Figure 5.7; see color section). Although it is tempting to consider that the oldest lunar basalts, enriched in both iron and titanium, were the most primitive and that younger basalts were consistently lower in titanium content, evidence shows that this was not the case. Lunar volcanic deposits appear to have come from various depths within the Moon, to have variable compositions at different times, and to have erupted in volcanic centers distributed mostly across the lunar near side.

## 5.5   DIFFERENCES FROM EARTH

Perhaps the major difference between volcanoes and their deposits on the Earth and Moon is their age. Volcanoes on Earth are active today—one can visit and study a number of erupting volcanoes on Earth at any given time. Volcanoes on the Moon are truly ancient, mostly ranging in age from 4.3 billion years to about 3 billion years. Even the youngest lava flows on the Moon are thought to be more than 1 billion years in age. For comparison, the oldest basalts on Earth are found in Greenland and they are only about 3.9 billion years old, while the basaltic floor of most of the Earth's oceans range in age from 180 million years to quite recent geologic time. These differences indicate that volcanic activity on Earth is an ongoing process, while the Moon appears to have "died" before complex multi-cellular life appeared on Earth.

Volcanoes are found in very different places on the Earth and Moon. On Earth, volcanoes tend to occur in linear chains, either adjacent to major mountain belts or marking the movement of crustal plates over a volcanically active zone ("hotspot"). The movement of crustal plates, or "plate tectonics", on Earth has clearly had a profound influence on the creation of magma (see Chapter 2). By contrast, lunar volcanic deposits are often circular because they fill most of the large impact craters on the Moon. Presumably this is because the impacts fractured the lunar crust and allowed melted rock beneath to flow upward. The observation that most lunar maria are on the near side of the Moon—the side that is always facing towards the Earth—has been explained as due to the presence of a thinner crust on the near side. Impacts on the far side of the Moon were unlikely to break through the thicker crust there, and indeed we see much less frequent evidence of volcanism on the far side of the moon. It is readily apparent that the process of plate tectonics has not played a role in volcanism on the Moon.

Finally, major environmental differences between the Earth and Moon have produced volcanoes and volcanic deposits that have different physical properties. First, lunar gravity is only about one-sixth that on the Earth, so volcanic materials are less likely to reach the surface on the Moon. This happens because the driving factors for magma ascent that are affected by planetary gravity, especially conduit width (affected in turn by the strength and density of surrounding rocks, depth of the source region, and local lithostatic pressure gradients), are reduced. Also, the atmosphere on the Moon is virtually non-existent, so volcanic materials were less affected by gravity and atmospheric drag and were thus able to travel substantially greater distances from their vents. For similar reasons, lava flows on the Moon were much more fluid and they often traveled much greater distances and spread out farther than their terrestrial counterparts. Thus we don't see either great mounds or mountains of volcanic deposits on the Moon. Second, the Moon is also a very dry place; gas and steam from water were not present to drive lunar eruptions, so they were much less violent and energetic than those on Earth. Lunar volcanic eruptions must have been spectacular to watch, but they were unlikely to have been the dramatic and sometimes long-lived events that we have seen on Earth.

## 5.6  RETURN TO THE MOON

These words may lead you to think that the story of volcanoes and volcanic deposits on the Moon has been told to completion. But that is not the case at all. There is much that we don't understand about the volcanic history of the Moon, and we have come to understand that it is important to our own health and perhaps future welfare that we understand our nearest neighbor in space as well as possible. First, we want to know everything about the Moon that helps us to understand what the Earth may have been like early in its history. Studies of the lunar impact craters and their temporal and spatial relationships with early lunar volcanic deposits will help us to understand the population and distribution of impacting bodies in the solar system that may have influenced the emergence and evolution of life on Earth in the past and may affect us all in the future. Although we know that volcanic activity on the Moon started early in its history and that volcanic eruptions occurred for something like 3 billion years, we still don't know the composition of the interior of the Moon and how it may have changed over time as it cooled. We don't really know how the maria erupted, and especially what their vents may have looked like. We see very few volcanic vents on the Moon, and we presume that this is because they were covered up by younger lava flows. Were the first volcanic vents huge, or were they small? Did they generally start with fizzy, fiery pyroclastic eruptions, or were they always quiet? How many lava flows did it take to fill up the lunar basins? Were there lots of thin flows, or a few thick flows? How much did impact cratering really have to do with where the earliest volcanic eruptions occurred on the Moon? These are just a few of the questions we have about volcanoes and their deposits on the Moon.

Ask any "lunatic" planetary scientist (as many of the early lunar scientists called themselves), and they'll tell you that we must return to the Moon. The Moon is a relatively simple planetary body, but it has been influenced by many of the same geologic processes that have shaped the Earth and the other terrestrial planets. If we don't understand the Moon, our nearest neighbor in space, then how can we understand the early Earth, or Mars, or Venus? By returning to the Moon, we can learn how to explore other planets and there we can begin to seek and find the answers to the big question of how we came to be as we are.

## 5.7  REFERENCES

Adams, J.B., Pieters, C., and McCord, T.B. (1974) Orange glass: Evidence for regional deposits of pyroclastic origin on the moon. *Proceedings of the 5th Lunar Planetary Science Conference*, pp. 171–186.

Delano, J.W. (1986) Pristine lunar glasses: Criteria, data, and implications (Proceedings of the 16th Lunar Planetary Science Conference, Pt. (2)). *J. Geophys. Res.*, **91**(Suppl.), D201–D213.

Gaddis, L.R., Pieters, C.M., and Hawke, B.R. (1985) Remote sensing of lunar pyroclastic mantling deposits. *Icarus*, **61**, 461–489.

Gaddis, L.R., Staid, M.I., Tyburczy, J.A., Hawke, B.R., and Petro, N. (2003) Compositions of lunar pyroclastic deposits. *Icarus*, **161**, 262–280.

Head, J.W. and Wilson, L. (1979) Alphonsus-type dark-halo craters: Morphology, morphometry, and eruption conditions. *Proceedings of the 10th Lunar Planetary Science Conference*, pp. 2861–2897.

Head, J.W. and Wilson, L. (1992) Lunar mare volcanism: Stratigraphy, eruption conditions, and the evolution of secondary crusts. *Geochimica et Cosmochimica Acta*, **56**, 2155–2175.

Heiken, G.H., Vaniman, D.T., and French, B.M. (eds) (1991) *The Lunar Sourcebook: A User's Guide to the moon.* Cambridge University Press, New York, 736 pp.

Hiesinger, H., Head, III, J.W., Wolf, U., Jaumann, R., and Neukum, G. (2003) Ages and stratigraphy of mare basalts in Oceanus Procellarum, Mare Nubium, Mare Cognitum, and Mare Insularum. *J. Geophys. Res.*, **108**, 5065, doi:10.1029/2002JE001985.

Neal, C.R. and Taylor, L.A. (1992) Petrogenesis of mare basalts: A record of lunar volcanism. *Geochimica et Cosmochimica Acta*, **56**, 2177–2211.

Robinson, M.S. and Lucey, P.G. (1997) Recalibrated Mariner 10 color mosaics: Implications for mercurian volcanism. *Science*, **275**, 197–200.

Shearer, C.K. and Papike, J.J. (1993) Basaltic magmatism on the moon: A perspective from volcanic picritic glass beads. *Geochimica et Cosmochimica Acta*, **57**, 4785–4812.

Spudis, P.D. (1996) *The Once and Future moon.* Smithsonian Institution Press, 308 pp.

Staid, M.I. and Pieters, C.M. (2001) The mineralogy of the last lunar basalts. *J. Geophys. Res.*, **106**, 27887–27900.

Wilhelms, D.E. (1987) The Geologic History of the Moon (Professional Paper 1348), US Geological Survey, Washington, DC, 302 pp.

# 6

# Volcanoes on Mars: The global view

*Susan Sakimoto* (Goddard Earth Sciences and Technology Center)

*I am a planetary geologist who studies volcanoes on different planets. Even when I was little, I used to stare out the windows on car trips, speculating on how the terrain we passed happened to get where it was, looking like it did. As night fell, I often thought about what other star systems might have for worlds, and wondered if they had, and if their residents took road trips as well. Somehow, since I had grown up in the northern Rocky Mountains, I never wondered if these hypothetical aliens had mountains or not . . . that seemed like a necessity to me! As I got older, I slowly realized that there was an entire field of people who got to think, for a job, about how landforms are created—geologists. Through elementary and middle school I remained interested in understanding "how the Earth works", but this interest got a good boost in late middle school, when Mount St. Helens erupted and scattered ash over my hometown in northern Idaho, closing the school, shutting down businesses, and generally making the volcano the sudden focus of everyday life. I thought that this was a pretty awesome display of how to change a terrain, and that it was something I really wanted to understand. I eventually went to college as a prospective geophysics major, and got diverted a bit by the astronomy department and their planetary surface images. I graduated with a Geology degree and minors in physics and astronomy, and went to graduate school for a masters and doctorate at Johns Hopkins University in Earth and Planetary Sciences. Volcanoes on other planets? Throughout my adolescent years (and even now) I have been totally fascinated by the similarities and differences of volcanoes in different environments. Nearly everywhere we look we find volcanism—our continents and sea floors, the Martian and Venusian surfaces, our Moon, and even the moons of the gas giants. The volcanic process can be seen with all kinds of local flavors and styles. Nowhere are any of them quite the same, but everywhere the landforms are (we think!) still recognizable as essentially the same processes. All these volcanoes not only look really cool, their endless differences raise all sorts of questions of whether one factor or another dominated their creation and emplacement, and whether we can tease out an answer from the data we have. I now work for the Goddard Earth Sciences and*

*Technology Center at the Geodynamics Branch, NASA Goddard Space Flight Center, and spend my time mostly looking and thinking about Martian and terrestrial volcanoes. I have two kids who sometimes join me in fieldwork on Earth, and I am married to an astrophysicist who luckily does not mind a house and yard full of assorted rocks and piles of spacecraft data and images, and enjoys "vacations" puttering about assorted volcanic landforms.*

Mars has a real treasure trove of volcanoes, and we have both landing and orbital instrument data for Mars. While I love to speculate on rock types and rock samples, I am happiest when my data for volcanoes shows their shape or topography, and thus most of my work is from orbital instruments. (See Chapter 7 for lander-based Mars investigations.) Mars is a beautiful planet to a volcanologist. It has, among other things, a wide range of volcanic features, including shield volcanoes, lava flow fields, and scores of small shields and volcanic fields. Mars hints that much of its history has included volcanism of some kind or another (see, e.g., Hodges and Moore, 1992; Francis, 1993; Cattermole, 1997). From orbit, we can look at a range of volcanoes from big to small, compare compositions (on a grand, planet-wide scale), provide context for lander samples, and study the topography, images, and compositional data of both large and small features for hints on how they might have been created.

## 6.1    TOPOGRAPHY FROM ORBIT

The Mars Orbiter Laser Altimeter (MOLA) instrument on the Mars Global Surveyor (MGS) spacecraft has provided global topography for Mars (e.g., Smith *et al.*, 1999), and it has been invaluable for studying the volcanoes. One of the most basic attributes we can measure with altimetry (the measurement of topography or elevations) is volcano size, characterized by height, diameter, or volume. Some of the larger single volcanoes in the solar system are located on Mars, just a (relatively) short hop from the Earth, and some are so big that they are easily seen on a global map (see Figure 6.1; color section).

Although Mars has hundreds of volcanoes; the largest naturally attract quite a lot of interest. One of the volcanoes with the largest volume is Alba Patera (see Figure 6.1; color section). Depending on the exact altitude defined as the base of the volcano, Alba Patera is 5–10 km high, but its slopes are shallow (0.1°–2°), and with a diameter of nearly 2,000 km, the total volume dwarfs any other volcano on the planet, including that of the taller, narrower Olympus Mons. The topography data for Alba suggests one explanation for the low slopes and impressive extent of Alba Patera: lava tubes are readily visible on the volcano's flanks. Lava tube flows are flows where the lava moves along a single or series of insulated flow paths within the lava flow (see Chapter 2). These covered conduits insulate the lava from some of the cooling that a sheet-like flow of similar size might encounter, and thus the lavas are able to move farther downhill, increasing the lateral extent of the volcano. Figure 6.2 shows a section of Alba Patera. The long ridges are thought to be lava

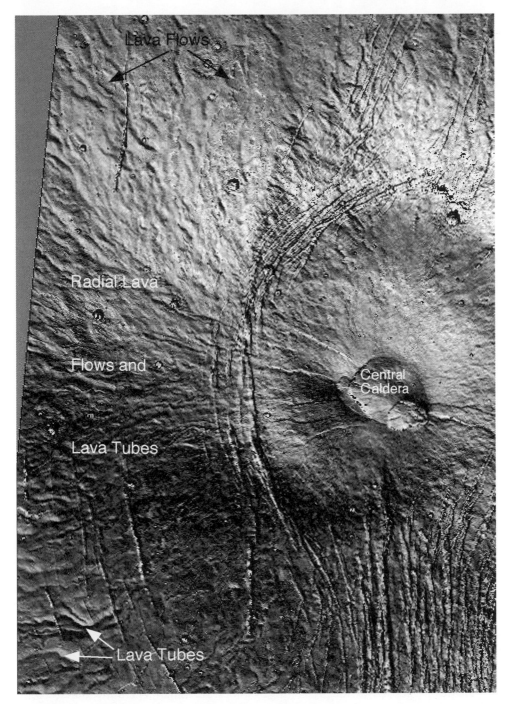

**Figure 6.2.** Topography from the MOLA for a section of Alba Patera showing long ridges thought to be lava tubes. The caldera is 110 km across.

tubes and several features add to our confidence that they are. First, the ridges always run downhill even at very low slopes (from a few degrees to a few hundredths of a degree). Second, some of them have small collapse pits along the ridge tops that reveal empty voids inside (just like terrestrial lava tube caves). Third, some of the ridges are the apparent sources of lava flows either along the ridge extent or at the ridge ends, evidence which supports that the flows are transported within the ridges.

Alba Patera is only a fraction of the height of Olympus Mons, the tallest volcano on Mars, which is about 24 km high and over 600 km across. Olympus Mons would dwarf either Mount Everest or Mauna Loa (the largest terrestrial volcano edifice) in both height or volume (see Figure 6.3; color section).

Most of the Martian volcanic features—basaltic volcanoes, scoria cones, lava flows, lava tubes, small shields, etc.—are much larger than their counterparts on Earth (e.g., Hodges and Moore, 1992; Cattermole, 1997). A lack of plate tectonics explains some of this. On Mars, the plates (and the volcanoes on top of them) apparently do not move across the mantle heat sources like they do on Earth (see Chapter 2), and thus a hot spot does not create a chain of volcanoes, but rather one large volcanic pile. Compare the Hawaiian chain in Figure 6.3 (see color section) with Olympus Mons. If you were to collect all of the Hawaiian Islands together, you actually start to approach a volume of material that would be noticeably comparable to Olympus Mons (although still smaller by at least one-half). Comparing Hawaii volcanoes to Olympus Mons and other Martian volcanoes is more reasonable than it might appear based on size alone though—the compositions are thought to be similar based upon their shapes (see, e.g., Carr and Greeley, 1980; Hodges and Moore, 1992; Francis, 1993).

Figure 6.1 (see color section) shows that Olympus Mons is one of a handful of volcanoes that are so tall that they are global landmarks. Olympus Mons, its three neighbors (Arsia, Pavonis, and Ascreaus Mons, all 10–15 km high), and the Elysium Mons volcanoes are the giants in height, and all of them have diameters in the hundreds of kilometers ranges. While they are high, they are much smaller in diameter than Alba Patera, and therefore do not have total volumes as large as Alba Patera. Their slopes are 3–5°, which is typical for terrestrial shield volcanoes, but perhaps a bit steeper than we would expect for a basaltic shield volcano on Mars. However, the Global Thermal Emission Data on the MGS spacecraft suggests that these edifices might have either a little more silica than a typical terrestrial basaltic shield (like Mauna Loa, Hawaii), or alternatively, be slightly water-altered. To tell the difference, we would have to send a lander to remove the surface dust from the volcanic rocks, and take a closer look at the volcanic rock compositions. The surface dust is thought to be the fallout of local, regional, and global dust storms and is probably from all over Mars (e.g., Christensen et al., 1998). When it falls to the ground, it coats local rocks, obscures their textures, and thus often hides their true compositions. Taking the spectra of a dust-covered region from orbit may give you a mix of dust and rock or surface compositions, or just the dust, if it is thick enough.

However, not all of the Martian volcanoes are huge. There are hundreds of volcanoes that are much smaller than the biggest few, and they have helped resurface a larger fraction of the planet as well (e.g., Garvin et al., 2000). Many of

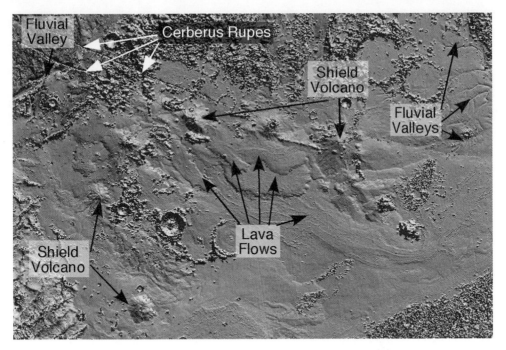

**Figure 6.4.** Topography for the Elysium Planitia and the Cerberus Rupes regions, showing extensive lava flows, small shields, and rift events. (Figure is approximately 1250 km across.)

these smaller shield volcanoes have shallow slopes (0.1 degree to several degrees), and very extensive but subtle flank slopes. The majority of them were essentially invisible until we obtained detailed topography of the planet, as they are not steep enough to be easily seen in images, and it is easy to be uncertain if the summit crater is an impact crater or part of a volcano if the volcanoes flanks are not readily apparent. However, when we can see the volcano shape and slopes in the topographic data, we can normally tell if something is a volcanic crater, or is simply another impact crater on a volcano's flanks. These smaller shields occur in large fields of shields, as part of vent structures in lava fields, as secondary features on top of larger volcanoes, and as widely scattered singles and small groups. Through the duration of the MGS mission, it has been very interesting to see more and more of the subtle topographic features become visible as the topographic grid resolution has improved—the small shields are a good example of this.

Smaller volcanic features like lava flows, lava tubes, and cratered cones (perhaps similar to terrestrial cinder cones), are also more visible in the topography than we might have anticipated. For example, the rather smooth plains of the Elysium Planitia and the Cerberus Rupes regions just north of the equator were rather enigmatic before the recent missions, and both water and volcanism had been suggested as processes for their origins (e.g., Carr, 1996). Figure 6.4 shows some of the MOLA topography for the Elysium Planitia and the Cerberus Rupes regions.

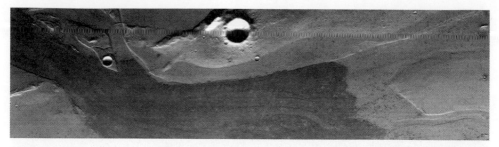

**Figure 6.5.** MOC image of Marte Valles. The empty river channel can be seen, which is partially filled by the dark lava flow that flowed down the channel. (Figure is approximately 41 km across.)

Lava flows make channels, as does water flowing down a slope. With topography, we can usually tell the lava channels from the water channels. For example, the lava channels build their own levees (e.g., Hughes *et al.*, 1999), and are often raised above the surface or make their own ridge running downslope. Water channels follow the low topography, do not build the same kind of levees, and are extremely unlikely to run down the center of a ridge. Some recent research suggests that lava and water are both using low valleys as flow paths and perhaps the same vent system as a route to the surface in Elysium (e.g., Burr *et al.*, 2002).

In the topography, extensive and fairly topographically fresh volcanic features are abundant, confirming that this region has been the site of repeated and geologi-cally young volcanic activity. While a few fluvial features are visible in Figure 6.3 (see color section), after close inspection it helps to have camera images as well for comparison. MGS's Mars Orbiter Camera (MOC) images showed young lava flows (tens of millions of years old) (Hartmann and Berman, 2000) flowing down somewhat older river channels (see Figure 6.5 and Malin and Edgett, 1999). These images and others like them show that the river channels were in place before the lavas erupted. We can use a combination of images like these and topography to determine which channels are lava channels, which are fluvial, and which are fluvial channels flooded with lava. The superposition of the lava and fluvial processes in this region make this an interesting area to consider for past life environments, as water and a heat source together raise the likelihood that the environment might have been hospitable to life forms (e.g., Gross, 1998; Taylor, 1999; Burr, 2002).

Topography alone, or topography combined with other data sets is contributing substantial additional insights into Martian volcanism processes.

## 6.2   IMAGES FROM ORBIT

The MOC images from the MGS spacecraft and the THermal EMission Imaging Spectrometer (THEMIS) images from the Mars Odyssey (MO) spacecraft are also revealing details of volcanic features previously unseen. The lava flow in Figure 6.5 is one good example. Figures 6.6 and 6.7 show additional examples. Figure 6.6 shows

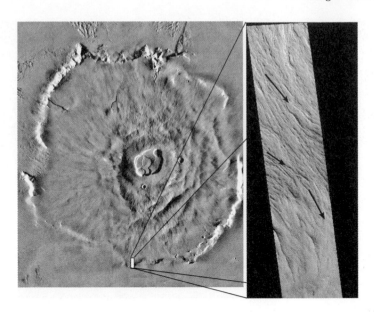

**Figure 6.6.** MOC image (right) of a portion of Olympus Mons (left). Extensive and complex lava flows with channels and levees easily visible. Olympus Mons is about 600 km in diameter.

details of lava flows on Olympus Mons, illustrating complex channel formation within the flows, and suggesting that flow emplacement here—in channels—is very different from that at Alba Patera—mostly in lava tubes.

MOC images of lava flows reveal distinct appearances, which suggests different flow emplacement styles. For a contrast to the narrow channel flows in Figure 6.6, see Figure 6.7. Both types of flows have levees, but in Figure 6.7 there is a thin flow over a much shallower slope (a few hundredths of a degree instead of a few degrees as in Figure 6.6). The flow in Figure 6.7 is much wider, and its flow surface shows a platy texture (a top surface apparently with plate-like crust pieces still preserved on it). Both the texture (observed in the image) and the slopes (measured from the topographic data) suggest to us that these Elysium Planitia flows were emplaced much more slowly than those at Olympus Mons.

In addition to flow details, MOC images of other features help us understand more of the complexity of Martian history. Figure 6.8 shows a portion of the Valles Marineris walls. The MOC camera team has suggested that the steep walls and visible layering observed all the way down to the canyon floor might be very similar to terrestrial flood basalts emplaced in layers on the Earth. If true, this would suggest quite a lot of basalt layers (more than 5 km thick if all the layers visible in the valley walls are basalts) for the region, which would be much more volcanism here than we previously suspected.

THEMIS images that are currently being returned from the Odyssey spacecraft have a slightly lower resolution than the MOC images (around 18 m/pixel instead of 2–10 m/pixel). The larger context of these images sometimes either helps us under-

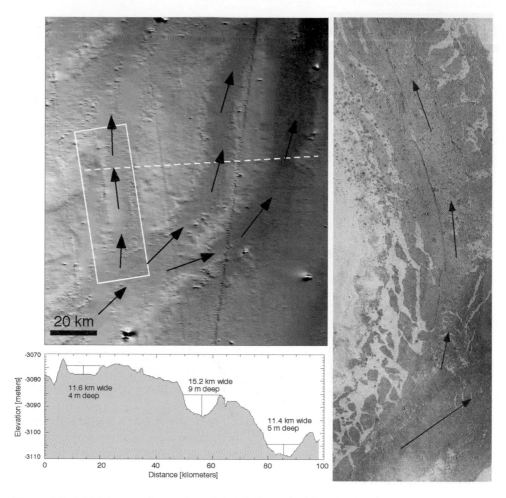

**Figure 6.7.** MOC image of a portion of the Elysium Planitia. The platy lava flow texture, wide extent, and shallow slopes suggest slower emplacement than the flows in Figure 6.6.

stand the higher resolution MOC images, or simply illustrates the volcanic nature of the surrounding terrain well. For example, the image in Figure 6.9 shows a lava flow inside an impact crater, and confirms that the (now) shallower crater became this shallow by being filled—at least partially—with lava flows. In areas with extensive lava flows we often see numerous impact craters with flat floors, and evidence for lava flows on the floors, or going through gaps in the crater rims. From looking at global measurements of impact craters, we know approximately how deep a fresh (little or no alteration) impact crater should be at each crater diameter (distance across the crater from rim to rim). So, we can compare a filled crater to the expected depth of a fresh crater of the same diameter, and estimate how much material in either volume or thickness has filled a crater. This is useful when we are estimating

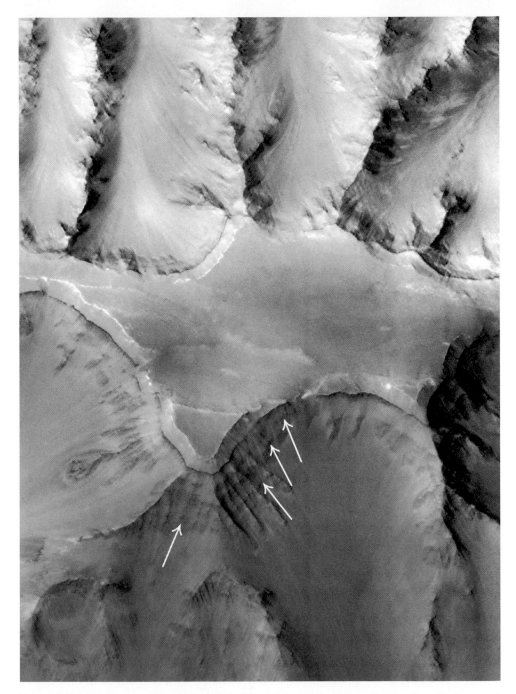

**Figure 6.8.** MOC image of a portion of the Valles Marineris. The walls show layering (arrows) somewhat like those seen in flood basalts all over the Earth, suggesting possible common origins. (Figure is approximately 15 km across.)

**Figure 6.9.** THEMIS image at 19 m/pixel of a lava flow within an impact crater in Daedalia Pannum (21.5°S, 229.7°E). North is towards the top. The impact crater is about 50 km across and portions of the rim can be seen at the top of the image.

surface ages or burial rates, and it can help determine which regions have had more volcanic activity.

Images such as the MOC and THEMIS images help us understand volcanic regions better by illustrating different kinds of volcanic emplacement, confirming or supporting the volcanic origins of regions, and raising new questions for further geologic investigations.

## 6.3 COMPOSITION FROM ORBIT

It is difficult to determine compositions of rocks from orbit. On Mars, the biggest problems are caused by the dusty atmosphere and the layer of dust found over every surface (with thicker layers often on older surfaces). This dust obscures the rocks (Christensen et al., 1998), as it does even for surface landers (see Chapter 7), and so our orbital data is often an indication of dust composition. The global data from the Thermal Emission Spectrometer (TES) on the Mars Global Surveyor spacecraft suggested that the materials in the southern parts of Mars appear to be basaltic, whereas those in the northern regions appear to be more silica rich (Bandfield, 2002). However, the same results might also be interpreted as showing that the northern region materials were water-altered basalt and those in the southern regions were less altered, but still basaltic. THEMIS has a higher resolution than TES, but data analyses are still hindered by the dust cover and ambiguity of silica and water alteration. However, THEMIS images have sufficiently high resolution that the thermal infrared images taken at night can help differentiate between hard rock areas and dustier areas, which helps significantly in geologic interpretation. As an additional compositional note, the MO Gamma Ray and Neutron Spectrometer measurements suggest that much of the polar and lower regions of Mars have abundant water ice in the pore space of the surface materials within the top few meters (e.g., Feldman et al., 2002; Mitrofanov et al., 2002). This has an effect on the weathering of the volcanic features as well as their emplacement (if the ice was there during their formation). This discovery of widespread water ice in the near surface and possibly at the surface (e.g., Mallin and Edgett, 2000) has invigorated discussions about where to search for possible life on Mars as well as discussion about exploration resources, since any landing crew would need water to survive.

## 6.4 DISCUSSION

Volcanism on Mars is diverse, and numerous types and sizes of lava channel, lava flow, volcano, and volcanic vent features can be seen and studied from orbital data and compared with terrestrial volcanic features. Higher resolution of Martian orbital data and the increasing diversity of available data types sometimes means we have better Martian data available than we do for terrestrial volcanoes, and stimulates collection of additional terrestrial data. However, it is much easier to reach terrestrial field sites to sample the volcanoes than it is ever likely to be to sample most Martian volcanoes, and so comparisons between terrestrial and Martian volcanoes, and orbital studies of the Martian volcanoes are likely to be our main mode of investigation of Martian volcanism for many years to come. Until we can return samples from Martian volcanoes, the topographic and image and spectroscopy data from orbit will allow us to constrain the types and durations of Martian eruptions somewhat. For example, comparing the types and dimensions of volcanic landforms will help us constrain how they were emplaced and over how much time, and comparing the kinds of landforms observed in different regions (e.g., cinder-like cones, channels, large sheet-like flows, lava tubes, etc.) allows us to compare the Martian landforms to known terrestrial volcanic landforms.

## 6.5　REFERENCES

Bandfield, J.L. (2002) Global mineral distribution on Mars. *J. Geophys. Res.*, **107**, 9–1, doi 10.1029/2001JE001510.

Burr, D.M., McEwen, A.S., and Sakimoto, S.E.H. (2002) Recent aqueous floods from the Cerberus Fossae, Mars. *Geophysical Research Letters*, **29**(1), 10.1029/2001GL013345.

Carr, M.H. (1996) *Water on Mars*. Oxford University Press, New York, 1229 pp.

Carr, M.H. and Greeley, R. (1980) *Volcanic Features of Hawaii: A Basis of Comparison for Mars*. NASA Scientific and Technical Information Branch, Washington, D.C., 211 pp.

Cattermole, P. (1997) *Planetary Volcanism: A Study of Volcanic Activity in the Solar System* (2nd edition). Ellis Horwood Ltd., Chichester, UK.

Christensen, P.R., Anderson, D.L., Chase, S.C., Clancy, R.T., Clark, R.N., Conrath, B.J., Kieffer, H.H., Kuzmin, R.O., Malin, M.C., Pearl, J.C. *et al.* (1998) Results from the Mars Global Surveyor Thermal Emission Spectrometer. *Science*, **279**(March 13), 1692–1698.

Feldman, W.C., Boynton, W.V., Tokar, R.L., Prettyman, T.H., Gasnault, O., Squyres, S.W., Elphic, R.C., Lawrence, D.J., Lawson, S.L., Maurice, S. *et al.* (2002) Global distribution of neutrons from Mars: Results from Mars Odyssey. *Science*, **297**(5 July), 75–78.

Francis, P.W. (1993) *Volcanoes: A Planetary Perspective*. Oxford University Press, Oxford, UK.

Garvin, J.B., Sakimoto, S.E.H., Frawley, J.J., Schnetzler, C.C., and Wright, H.M. (2000) Topographic evidence for geologically recent near-polar volcanism on Mars. *Icarus*, **145**(June), 648–665.

Gross, M. (1998) *Life on the Edge: Amazing Creatures Thriving in Extreme Environments*. Plenum Press, New York, 250 pp.

Hartmann, W.K. and Berman, D.C. (2000) Elysium Planitia lava flows: Crater count chronology and geological implications. *J. Geophys. Res.*, **105**, 15011–15026.

Hodges, C.A. and Moore, H.J. (1992) *Atlas of Volcanic Landforms on Mars* (Professional Paper 1534). US Geological Survey, Reston, VA, 194 pp.

Hughes, S.S., Smith, R.P., Hackett, W.R., and Anderson, S.R. (1999) Mafic volcanism and environmental geology of the eastern Snake River Plain. In: S.S. Hughes and G.D. Thackray (eds), *Guidebook to the Geology of Eastern Idaho*. Idaho Museum of Natural History, ID, pp. 143–168.

Malin, M.C. and Edgett, K.S. (1999) MGS MOC The first year: Geomorphic processes and landforms. *Lunar and Planetary Science Conference (LPSC XXX), Houston, Texas*, CD-ROM, Abstract #1028.

Malin, M.C. and Edgett, K.S. (2000) Evidence for recent groundwater seepage and surface runoff on Mars. *Science*, **288**, 2330–2335.

Mitrofanov, I., Anfimov, D., Kozyrev, A., Litvak, M., Sanin, A., Tret'yakov, V., Krylov, A., Shvetsov, V., Boynton, W., Shinohara, C. *et al.* (2002) Maps of subsurface hydrogen from the High Energy Neutron Detector, Mars Odyssey. *Science*, **297**(5 July), 78–81.

Smith, D.E., Zuber, M.T., Solomon, S.C., Phillips, R.J., Head, J.W., Garvin, J.G., Banerdt, B., Muhleman, D.O., Pettengill, G.H., Neumann, G.A. *et al.* (1999) The global topography of Mars and implications for surface evolution. *Science*, **284**(5419), 1495–1503.

Taylor, M.R. (1999) *Dark Life: Martian Nanobacteria, Rock-Eating Cave Bugs, and Other Extreme Organisms of Inner Earth and Outer Space*. Scribner, New York, 288 pp.

**Additional bibliography**

Chapman, M.G., Allen, C.C., Dudmundsson, M.T., Gulick, V.C., Jakobsson, S.P., Lucchitta, B.K., Skilling, I.P., and Waitt, R.B. (2000) Volcanism and ice interactions on Earth and Mars. In: J.R. Zimbelman and T.K.P. Gregg (eds), *Environmental Effects on Volcanic Eruptions: From Deep Oceans to Deep Space*. Kluwer Academic/Plenum Publishers, New York, pp. 39–73.

MacDonald, G.A., Abbott, A.T., and Peterson, F.L. (1983) *Volcanoes in the Sea: The Geology of Hawaii*. University of Hawaii Press, Honolulu, HI, 517 pp.

# 7

# Volcanoes on Mars: A view from the surface

*Joy Crisp* (Jet Propulsion Laboratory, California Institute of Technology)

*As a teen, I was most interested in mathematics, literature, and science, and keen on devouring books and frequenting libraries. In my second year as an undergraduate at Carleton College in Minnesota, I happened to take an introductory geology class and I found it more interesting, challenging, and fun than anything else I had tried, so I continued taking classes in geology and became hooked on it. A Carleton research project on interpreting the mineralogy and volcanic textures in Precambrian rocks was my first introduction to the wonderful world of being a "detective volcanologist". After getting an undergraduate degree in geology, I went to Princeton University. There, I got a PhD in geology, focusing my research on some ash flow and lava flow deposits in the Canary Islands and a compilation of rates of magmatism on the Earth. Next, I had a postdoctoral research position at the University of California, Los Angeles, running experiments to "cook" rocks at high temperatures and pressures until they melted and recrystallized, to better understand the pressure and temperature stability conditions for some of the minerals in the Canary Island magmas. Ever since then, I have been working at the Jet Propulsion Laboratory in Pasadena, California, where as a Research Scientist, I have studied the cooling and crystallization of lava flows on the Earth and Mars, theoretical modelling of lava flow dynamics, and infrared remote sensing of lava flows and eruption clouds on the Earth. I have had an exciting time working on the Mars Pathfinder mission as an Investigation Scientist and on the Mars Exploration rover mission as the Project Scientist. Interpreting the images, geochemistry, and mineral information from these Mars missions is more challenging than Earth-based volcanology, but a real thrill because of the exotic location!*

## 7.1  MARS SURFACE MATERIALS

What kinds of rocks are on the surface of Mars? Several of the meteorites from Mars we have found on Earth are volcanic or near-surface intrusive basalt. Others are ultramafic intrusive rock, rich in olivine and pyroxene. All of these meteorite rocks from Mars have been thoroughly analyzed, in terms of their mineralogy, chemistry, and other rock characteristics, but we don't have any geologic context for them because we don't know exactly where on Mars they came from. The crystallization ages of the Martian meteorites span a wide range from about 4.5 billion to 165 million years,[1] attesting to the long volcanic history of Mars. Although there is a preponderance of volcanic deposits on the surface of Mars (Greeley and Spudis, 1981; Hodges and Moore, 1994), there is also photogeologic evidence for sedimentary water and wind-lain deposits, crater impact deposits, and possibly hydrothermal alteration deposits.

Besides the rocks, the other major component of the surface of Mars is the yellowish brown soil-like material and dust (Maki *et al.*, 1999), which constantly shifts around as dust devils and dust storms come through. We still don't know how and when that orangey oxidized material formed, but one of the popular hypotheses being tested by scientists is that volcanic basalt reacted with water long ago on Mars, and was converted into a material similar to what is called palagonite on the Earth. Figure 7.1 (see color section) shows an example of palagonite formed on the Earth by the interaction of volcanic cinder with water, which results in a yellowish brown deposit. So, even the yellowish brown dirt on Mars is interesting to a volcanologist, because it likely began as a volcanic rock before it was oxidized. In some palagonites on the Earth, some of the original volcanic crystals remain intact. The Martian soil may still contain clues to the nature of its parent volcanic rock.

## 7.2   WORKING ON THE MARS PATHFINDER MISSION

Having specialized in the study of volcanism on the Earth, I never thought I would find myself studying Martian volcanism at the Jet Propulsion Laboratory (JPL). But somehow I did. Mars is a volcanologist's paradise, an exotic location chock-full of lava flows, cones, domes, and gigantic volcanoes, which are obvious in the images acquired from orbiting spacecraft (see Chapter 6). After landing a job at JPL, I carried out a research study of the crystallinity variations in a Hawaiian lava flow. However, I gradually became interested in what my JPL colleagues were working on, which was volcanoes on Mars, and ended up writing a NASA proposal to do my own research on Martian lava flows. In early 1994, the Mars Pathfinder project was looking to hire someone at JPL who was knowledgeable in Martian geology and the chemistry and mineralogy of rocks, to look out for the interests of the Alpha Proton X-ray Spectrometer (APXS) science investigation. I took the job as the APXS investigation scientist.

[1] Preferred radiometric crystallization ages, which are deduced from a careful examination of K–Ar, $^{39}$Ar–$^{40}$Ar, Rb–Sr, Sm–Nd, and U–Th–Pb ages, range from 4.5 billion years old (ALH48001), to 1.3–1.35 billion years (Chassigny and the nakhlite meteorites), to 165–475 million years for the shergottites (Nyquist *et al.*, 2001).

**Figure 2.9.** (a) Fountaining episode of Kilauea volcano, July, 1984. The fountain is almost 300 m high

**Figure 2.11.** (a) Open channel lava flows on Kilauea volcano. (b) Close-up photograph of aa flow front.
(a) USGS photograph by J. Griggs.
Modified from Kauahikaua *et al.* (2003).

**Figure 2.13.** (a) An "open" lava tube with lava flowing well beneath the solidified crust.

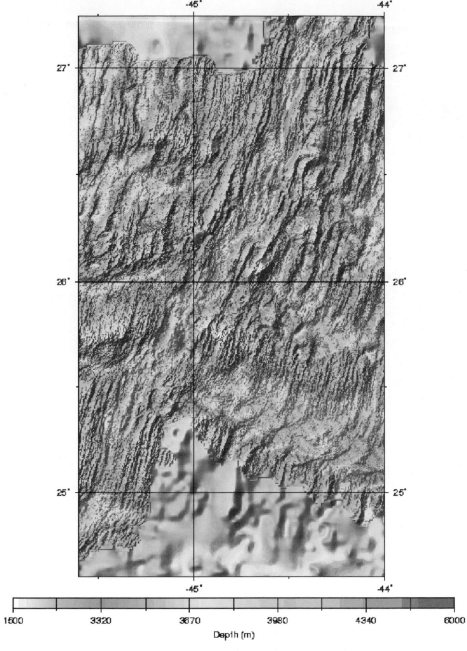

**Figure 3.8.** A bathymetric map of the Mid-Atlantic Ridge between 25°N and 27°N. The blue region trending diagonally across the map (from the lower left to the upper right) is the rift valley. The yellow hills within the blue rift valley are volcanoes—similar in shape to the shield volcanoes of Hawaii (see Chapter 2).

Map courtesy of the Ridge Multibeam Synthesis Project of Lamont–Doherty Earth Observatory.

**Figure 4.1.** This colorized image of Venus was taken 14 February, 1990 by the Galileo spacecraft on its way to Jupiter. The image was taken from a distance of almost 1.7 million miles, about 6 days after Galileo's closest approach to Venus. It has been image processed to a bluish hue to emphasize the details of the cloud markings. The color also indicates that it was taken through a violet filter to show features in the sulfuric acid clouds near the top of the planet's atmosphere that are most noticeable in violet and ultraviolet light. The smallest features visible are about 45 miles across. The Galileo Project was managed for NASA by JPL; its mission was to study Jupiter and its satellites and magnetosphere after multiple gravity-assist fly-bys at Venus and Earth.

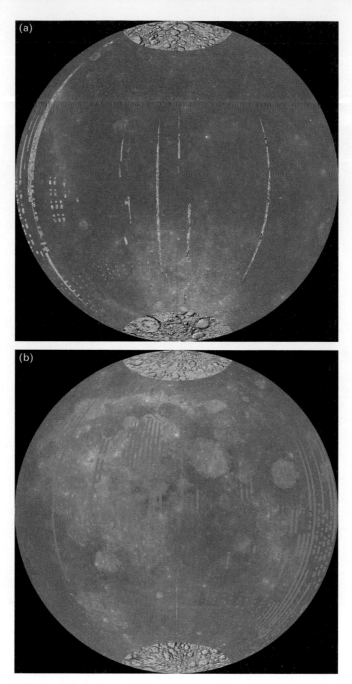

**Figure 5.7.** A "false color" view of the full Moon from Clementine made from ratios of images obtained at different infrared wavelengths. These mosaics show a colorized view in which red = 750 nm/415 nm, green = 750 nm/950 nm and blue = 415 nm/750 nm. The brilliant color variations in the lunar maria are due to compositional and age differences among lava flows, and these differences are easily seen when the Clementine images are viewed as color ratios. The lunar highlands are shown in shades of red (older) and blue (younger), while the lunar maria are shown in shades of yellow–orange (iron-rich, low-titanium content) and blue (iron-rich, high-titanium content). Superimposed on these units are materials from basins and craters of various ages ranging from ancient basins (red, dark blue) to young, fresh craters (bright blue). The Clementine data are superimposed on a shaded relief view of the Moon for clarity. Vertical stripes in the Clementine image represent areas of missing data in the mosaic. (a) Far side, Lambert Azimuthal Equal Area projection, centered at 180° longitude. (b) Near side, Lambert Azimuthal Equal Area projection, centered at 0° longitude. Courtesy of Matthew Staid, USGS.

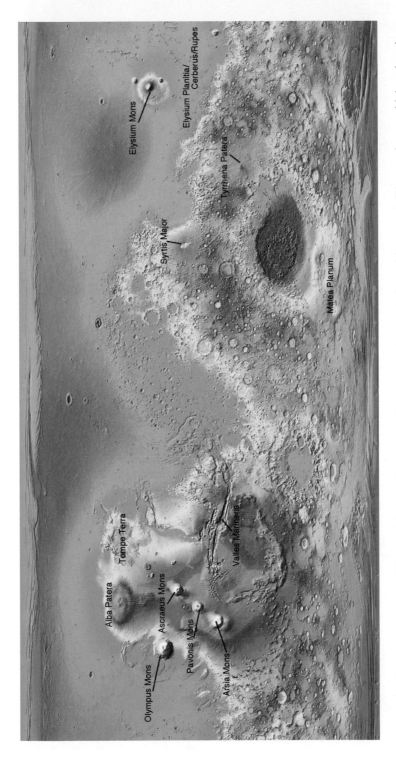

**Figure 6.1.** Topographic map of Mars (global), with several of the largest volcanoes indicated by arrows. Warm colors are higher elevations, and cool colors are lower elevations. Data is from MOLA on the MGS spacecraft. The larger volcanoes visible here are hundreds of kilometers across. The width of this image is approximately 21,600 km at the equator.

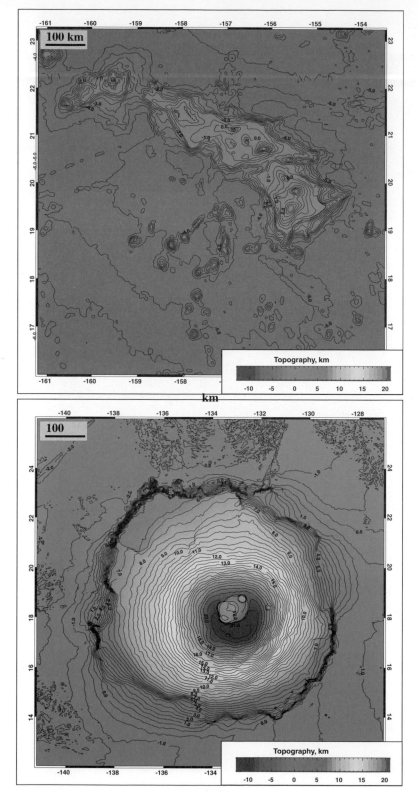

**Figure 6.3.** A comparison of the topographic maps of Olympus Mons, Mars (lower panel), and the Hawaiian Island chain on Earth (upper panel). The color and distance scales are the same for both.

**Figure 7.1.** Layers of palagonitized hydromagmatic ash with clasts of coral, pre-existing basalt, and basaltic lapilli at Koko Crater, Koolau volcano, Oahu.
Photo taken by Scott Rowland, University of Hawaii.

**Figure 7.4.** Color image of the Sojourner rover latched to the lander, taken by the Imager for Mars Pathfinder on 4 July, 1997. The image has been corrected for parallax curvature. Courtesy of NASA/JPL/Caltech.

**Figure 7.10.** Computer simulation by Maas Digital LLC, showing what one of the Mars Exploration Rovers could look like on Mars. For scale, the wheel diameter is 26 cm.

**Figure 8.1.** This beautiful image of Io was taken by Galileo's camera when Io was in front of Jupiter's cloudy atmosphere on 7 September, 1996. The colors in the image (a combination of the camera's near-infrared, green, and violet filters) have been enhanced to emphasize the vivid variations in color and brightness that characterize Io's volcanic surface. The black and bright red materials correspond to the most recent volcanic deposits and many of the black areas are active lavas. The near-infrared filter makes Jupiter's atmosphere look blue. The active volcano Prometheus is seen near the right-center of the disk. Part of the red deposit from the Pele plume is seen on the far left of the disk.

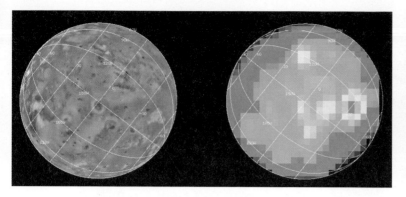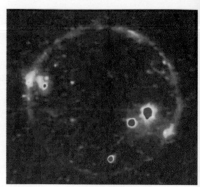

**Figure 8.5.** Galileo's NIMS and camera (solid state imaging system) detected numerous previously unknown hot spots on Io from distant observations before the fly-bys. A NIMS image (middle) at 4.1 μm shows several hot spots (in yellows and reds). A voyager image is shown for comparison. Galileo's camera was less sensitive to heat but could detect bright hot spots (in red) if it observed Io in total darkness during eclipses by Jupiter (right). The brightest hot spot in this image is Pillan (largest red area), during its raging eruption of 1997.

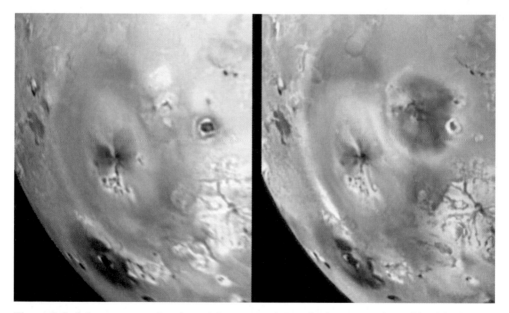

**Figure 8.6.** Io's constant volcanic activity causes dramatic changes such as this "black eye", the result of an eruption from the Pillan volcano in 1997. Pillan is located on the north-eastern part of the Pele red deposit. The changes, captured by the solid state imaging (CCD) system, occurred between the time Galileo acquired the left frame (4 April, 1997) and the right frame (19 September, 1997). The new dark spot, 400 km in diameter (roughly the size of Arizona) surrounds Pillan Patera and is thought to be ash and other pyroclastic deposits from a 120 km (75 mile) high plume. This eruption of Pillan was particularly important because it revealed that at least this Ionian volcano has extremely hot lavas.

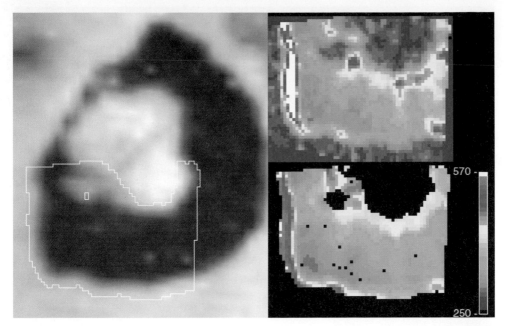

**Figure 8.9.** Loki, Io's most powerful volcano, is a 200 km wide caldera that contains a lava lake (dark areas in the image on left) surrounding a cold, light-colored island or topographic high. A NIMS map at 2.5 μm (top right) clearly shows a hot edge (indicated by white and red areas) next to the western caldera wall. A temperature map made from the same observation (lower right) shows that the highest temperatures are found near the caldera walls and against the walls of the cold island. The temperature scale is in Kelvin (570 K = 297°C). The approximate outline of the area observed by NIMS is shown in the camera image on the left. Modified from Lopes *et al.* (2004)

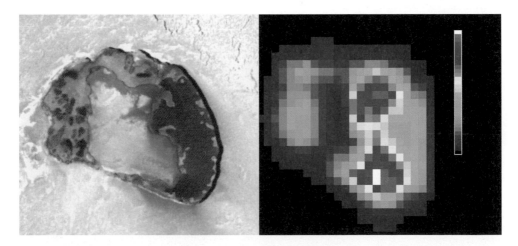

**Figure 8.10.** Tupan Caldera as seen by Galileo's camera (left) and infrared instrument (right) in October 2001. The infrared image uses false color to indicate intensity of glowing at a wavelength of 4.7 μm. Reds and yellows indicate hotter regions; blues are cold. The hottest areas correspond to the dark portions in the visible-light image and are probably hot lavas, probably forming a lava lake in the caldera. The central, cold region in the caldera may be an island or a topographically high region. Parts of it are cold enough for $SO_2$ to condense. Tupan, an active volcano on Io since at least 1996, was named after the Brazilian native god of thunder. The Tupan Caldera is about 70 km across.

**Figure 11.1.** Old Faithful Geyser in eruption. The plume is a mixture of hot upgoing water (in the center of the plume, not visible), a sheath of mixed-temperature water falling back toward the ground (the whitest core), and steam drifting away to the left and top. The center of the plume is about 30 m in height. The foreground is silica sinter, which is actually light white, but dark because of the enhancement in this photo. You can view eruptions on the Old Faithful WebCam (`http://www.nps.gov/yell/old faithfulcam.htm`). You can find out more about geysers by logging onto `www.geyser.com`, a website about geysers developed by the author.

Photo by S. Kieffer.

**Figure 11.2.** A geyser-like eruption from the summit of Mount St. Helens on 30 March, 1980. Steam and crushed rock are being ejected from the developing summit crater, rising on the left (east) to form a stubby dark vertical plume about 100 m high. Steam separates buoyantly from this dark area, appearing as whiter portions of the plume. The ejecta fall back (like the sheath on Old Faithful Geyser in Figure 11.1) and the fallback material is so dense that it flows downhill to the right (west) forming a small pyroclastic flow that emerges at the base of the mountain on the west. The Toutle River valley is in the foreground.
Photo by S. Kieffer.

**Figure 11.16.** "Geyser with Rainbow", a vision of the surface of Titan, by Mark Robertson, Tessic, and Ralph Lorenz. Droplets of methane-rich liquid may be erupted into Titan's atmosphere from methane geysers or fumaroles. Methane drops will make rainbows, which will appear red with a bit of yellow because high altitude organic haze absorbs blue light. Saturn is visible in the sky.

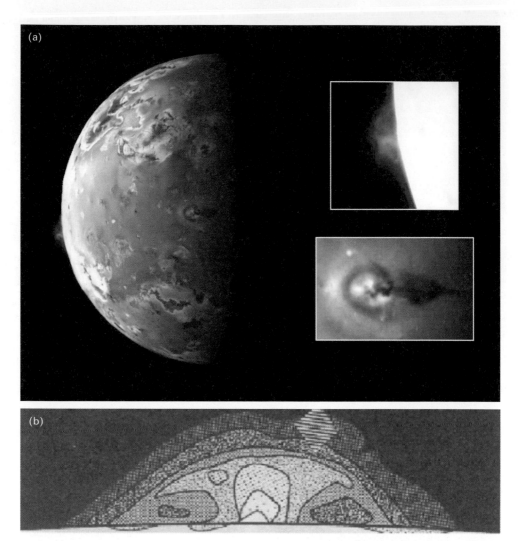

**Figure 11.6.** Plumes on Io. (a) The plume visible just below the center on the bright limb is enlarged in the top inset. It is erupting over a caldera named Pillan Patera after a South American god of thunder, fire, and volcanoes. It is 140 km high. The second plume visible just to the left of the terminator in the center of the photo, is shown in the lower inset. It is called Prometheus, after the Greek fire god. The plume is 75 km high, and its shadow can be seen extending to the right of the eruption vent; for a description of its behavior over 20 years since its discovery, see Kieffer *et al.* (2000). Prometheus has probably been continuously active since it was first observed in 1979; the plume at Pillan Patera was not seen before the Galileo spacecraft observed it in 1997. (b) Brightness contours at Prometheus in 1979.

(a) NASA/JPL-Caltech. (b) From Strom and Schneider, 1982, with permission.

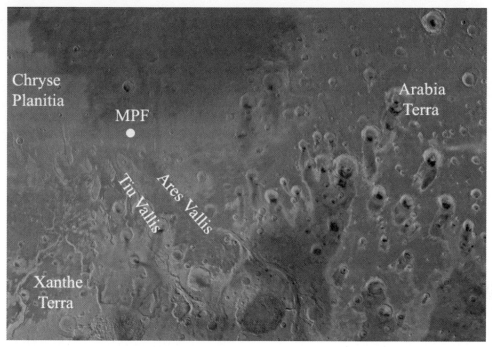

**Figure 7.2.** Viking Orbiter image mosaic PIA00171 covering the region of Mars from 0 to 30°N latitude and 0 to 45°W longitude. The white dot labeled "MFP" shows the location of the Pathfinder landing site.
Courtesy of NASA/JPL/Caltech.

One of the more interesting parts of the work was participating in the dilemma of where to send the Pathfinder lander. We had to use images and other orbital data sets to characterize the safety and science criteria of the site before we got there (Golombek *et al.*, 1997). It is challenging to select a site that is sufficiently safe for a spacecraft to land, yet is also scientifically interesting. One of the primary reasons for selecting the Pathfinder landing site was its location—in the path of an ancient catastrophic flood (Golombek *et al.*, 1997; Greeley *et al.*, 1977). The site is 100 km downstream from where the Ares Vallis channel widens out onto an open plain (Figure 7.2). Upon widening, the flood waters should have slowed considerably and dropped out any giant boulders it was carrying, leaving a much smoother terrain further downstream at the landing site.

It was hoped that this site might provide our lander camera and rover with a "grab bag" assortment of rock types that had been carried by floodwaters from the ancient terrain to the south. We thought we could learn more about the geologic history of Mars at this site, than at a site where there might only be one type of rock. From orbital spacecraft data, we weren't sure what might be in that "grab bag", but volcanic rocks were certainly a possibility. Sedimentary rocks with clues to Mars' past climate were also possible, as well as impact crater deposits. By sending a

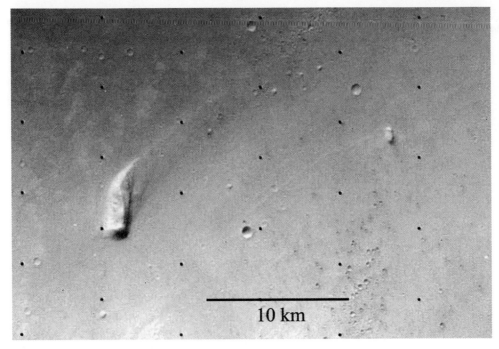

**Figure 7.3.** Cropped portion of monochrome Viking Orbiter 1 image 004a27 approximately 60 km from the Pathfinder landing site. Streamlined island indicates water erosion. The grid of black dots are reseau marks from geometric calibration.
Courtesy of NASA/JPL/Caltech.

mission to the surface to look close-up at the rocks and to examine them on the scale of pebbles to boulders, we had a chance of discovering what kind of rocks were there and how they formed. If the rocks had been transported from the ancient highlands, then we might learn something about the climate of Mars in the early part of its history.

The highest resolution images we had of the landing site from orbit were of the order of 40 m/pixel, which meant that we could just barely make out features that were 120 m in size! We couldn't be sure what we would find once we landed and saw things close-up on the surface. What was clearly visible in the orbital images were things like streamlined islands and knobs and erosional terraces (Figure 7.3), in alignment with the channels to the south that appeared to have been cut by ancient floodwaters. For Pathfinder, our navigational accuracy was such that we knew we would land somewhere inside an ellipse that was about 200 km by 70 km. About 110 km to the east, there were some landforms resembling low shield volcanoes that looked worn down, giving them a mottled appearance. This mottled terrain extended into the landing ellipse, even to within 5 km of the landing location, giving us volcanologists some hope we might encounter volcanic rocks.

The Martian day (called a "sol") is approximately 24 hours and 39.5 minutes long. The Mars Pathfinder project divided this up into three overlapping 10-hour work shifts, with the first shift examining downlinked images and data returned from Mars, and early and late uplink shifts preparing a new set of commands for the rover and lander every sol. I was assigned to the first uplink shift, so two days before the landing, I stayed up for 24 hours to adjust to my Mars schedule. I forced myself to sleep through Pathfinder's landing. I'm not sure how I did that—I must have been dreaming about Mars. My husband, who was an atmospheric scientist on the mission, woke up in the middle of the "day", checked the TV, told me that it landed okay, and we both went back to sleep. Then when I woke up at 5:30 pm for my first day of preparing uplink for Mars on 4 July, 1997, I saw the first image mosaic from Pathfinder on TV, and was amazed. There it was, the rover fully intact, surrounded by rocks and "Twin Peaks" in the background providing an interesting landscape (Figure 7.4; see color section)!

Then, the real challenge began. Because we didn't know how long the mission would last and we wanted to make the most of every day, we were under severe time pressure each day to learn what we could about the rocks and soil from the images and data that came back from Mars, and to come up with the best set of commands for the rover and lander to maximize our science return. Since the Mars clock is out of sync with the time on Earth, every day my work shift moved about 40 minutes forward, so that by 21 July, I was waking up at my usual time, 5 am. The "jet lag" effect was getting tiresome, although it was offset by the adrenaline rush from seeing new pictures come back from Mars and watching the rover drive to new locations. After the first 30 days of the landed mission, the project went back to working Earth daylight hours, which was a relief.

The geology science tools sent to Mars on this mission were quite limited. The lander was equipped with stereo cameras (Smith *et al.*, 1997), with different filters that could be rotated into position to capture different color images from blue through green, red, and infrared. The color information was used to map out different kinds of soils and rocks at the landing site and to put constraints on the kinds of iron-bearing minerals present. The image information was used for characterizing the morphology, structures, and textures of the surface materials at the local site, and stereo image pairs allowed us to calculate sizes and distances of features. The rover (Rover Team, 1997) had stereo cameras at the front, a color camera at the back, and an APXS for measuring the chemistry of rocks and soils (see Figure 7.5). From the lander camera images, we made decisions as to which rocks to send the rover to, for up-close examination. The rover's workday for driving and imaging was fairly short, but it could take 8 to 12-hour duration chemistry measurements during the night. Although it had a lot of limitations in its capabilities, the mission lasted longer than expected. Over its 83-day lifetime, the rover put 104 m on its "odometer" (see Figure 7.6) and we received 16,500 images from the lander camera ($14°$ field of view, $256 \times 248$ pixel images), 550 images from the rover ($41° \times 22°$ field of view, $768 \times 484$ pixel images), measurements from 14 soil mechanics experiments, and APXS chemistry measurements for 5 rocks and 6 soils. The rock and soil targets for chemical analysis were carefully selected to be

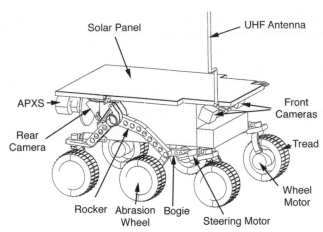

**Figure 7.5.** Line drawing of the Sojourner rover, pointing out various components. (Wheel diameter = 13 cm.)

representative of the range of materials at the landing site and sometimes required a few days to get the rover properly aligned to get a good deployment of the instrument onto the intended target. Figure 7.7 shows an example of a good deployment of the APXS on the rock called "Half Dome". See the Appendix at the end of this chapter for more details on the Pathfinder payload used for geology investigations.

## 7.3  WHAT DID MARS PATHFINDER TELL US ABOUT VOLCANISM ON MARS?

The main new discovery about the rocks on Mars was the higher than expected silica ($SiO_2$) content, which was in the basaltic andesite (52–57%) range (Rieder *et al.*, 1997). Previous to Pathfinder, most of the rocks on Mars were thought to be basaltic or ultramafic (less than 52% $SiO_2$). Viking images showed abundant features like lava tubes, shield volcanoes, and flood lavas that resembled basaltic volcanism on Earth (see Chapter 6). Recently updated calibrations indicate that the least dusty rock measured by the APXS has about 54–55% $SiO_2$ (Economou, 2001; Wänke *et al.*, 2001). If all of the measured sulfur in Pathfinder rock analyses can be attributed to the dust coatings on the rocks, and assuming the dust coatings have the same sulfur content as the soils, then the underlying rock probably has about 56–57% $SiO_2$. This is quite different from Martian meteorites, which have 37–53% $SiO_2$. Silica content of volcanic rocks is important because it is an indicator of how runny or sticky the lava was, the style of eruption (explosive or effusive), and how changed in chemistry it may have been from a "parent" magma. After the Pathfinder mission, an instrument on the orbiting Mars Global Surveyor called the Thermal Emission Spectrometer (TES) found evidence (mineral abundance estimates deconvolved from infrared spectra) that the rocks in the southern parts of Mars appear to

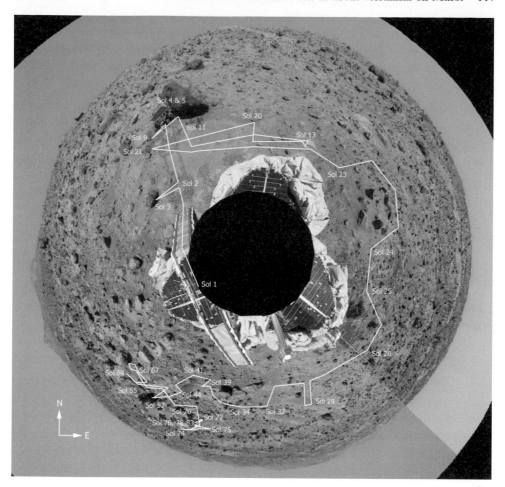

**Figure 7.6.** Azimuth–elevation projection of the panorama aquired at the Pathfinder landing site. Also shown is the driving path taken by the Sojourner rover during its surface mission. Courtesy of NASA/JPL/Caltech.

be mostly basaltic, whereas in the northern parts, including the Pathfinder landing site, they are either mostly andesitic in composition or water-altered basalt (Bandfield *et al.*, 2000; Wyatt and McSween, 2002). More details on this instrument are provided in the Appendix at the back of this chapter.

We had hoped that by looking close-up at rocks with the Pathfinder cameras, we would recognize textures that were diagnostic of how the rocks formed. If we had seen well-formed crystal shapes in a matrix of smaller ones, reminiscent of volcanic rocks on Earth, that would have been good evidence supporting a volcanic origin. However, the rocks were so coated with dust and affected by wind abrasion, that we were unable to see the underlying rock texture (McSween *et al.*, 1999). Some of the rocks exhibit faint lineations (see lower third of the rock in Figure 7.8), but we are

**Figure 7.7.** Sojourner rover with its APXS deployed against the rock called "Moe" on 7 September, 1997. This is a mosaic of two image frames acquired by the Imager for Mars Pathfinder.
Courtesy of NASA/JPL/Caltech.

**Figure 7.8.** This image of the rock named "Chimp" was acquired by the Sojourner rover's right front camera on 17 September, 1997. Lineations and pits are seen on the rock's surface. The rock is approximately 40 cm tall.
Courtesy of NASA/JPL/Caltech.

**Figure 7.9.** This image of the rock named "Stimpy" was acquired by the Sojourner rover's left front camera on 13 September, 1997. The black horizontal bar represents some rows of lost data. The rock is approximately 25 cm tall.
Courtesy of NASA/JPL/Caltech.

not sure whether they are indicative of sedimentary layers, wind-etched features, variations in volcanic crystallinity, or volcanic flow-banding. The holes in some of the rocks might be volcanic bubble holes called vesicles, but an alternative origin as wind-carved flutes cannot be ruled out (see Figure 7.9). One of the rocks resembled a sedimentary conglomerate in texture, but it was vague enough that other possibilities proposed included a volcanic agglomerate or impact breccia. We weren't even sure whether there was a true "grab bag" of different rock types present. The high silica content also remained a mystery, explainable as a weathered rind, impact melt, sedimentary deposit, or volcanic composition (McSween *et al.*, 1999). Unfortunately, we were left scratching our heads over the origin of the rocks at the Pathfinder landing site.

In addition to the ambiguous rock textures, the local landforms did not reveal anything diagnostic with respect to volcanism. Twin Peaks appear to be 50 m tall hills that were streamlined by the floods, rather than formed as volcanic cones (800–900 m from the landing site, on the horizon in the middle of Figure 7.4). The nearby craters are definitely impact craters, and not volcanic. Some of the boulders were lined up all in a row, leaning against each other at an angle that resembled flood deposition imbrication. But no volcanic lava flow structures or vents were seen at the landing site. The lesson here is that you don't always answer the questions you originally set out to in a planetary mission, but Pathfinder did improve our understanding of Mars and raise new questions.

On Earth, geologists walk around, observe, take notes, recollect similar things they have seen or read about, formulate hypotheses and test them, map contacts between different rock types, hammer off pieces of rock and examine them closely with a magnifying hand lens, bring samples back to the laboratory, and slice and polish rock slabs for microscopic examination. This may happen on Mars some day far in the future, but for now, geologists who study Mars have to be content to rely on robotic spacecraft to do their geology for them. We study the precious bits of data we get back from Mars every way we can, to put together the clues as to how the rocks formed, and what geologic processes were active. Even though we're not there on the planet, most of us feel like we have been there, because we study the images so intensely. And even though Mars may no longer be volcanically active, the volcanic features left behind are so astounding, that many of us remain captivated by the study of bits and bytes from Mars.

## 7.4   THE MARS EXPLORATION ROVERS AND OTHER FUTURE MISSIONS

The next NASA mission to Mars, the Mars Exploration Rover mission (Crisp *et al.*, 2003; Squyres *et al.*, 2003), landed two rovers named Spirit and Opportunity in two different locations on 4 January and 25 January, 2004, UTC time (Figure 7.10; see color section). These rovers are much more capable than the Pathfinder rover, and are outfitted with a Rock Abrasion Tool (similar function to a geologist's hammer), Microscopic Imager (similar to a geologist's hand lens), improved APXS, Mössbauer Spectrometer for identification of iron-bearing minerals and iron oxidation level, stereo Panoramic Camera, Miniature Thermal Emission Spectrometer, and Navigation and Hazard cameras. For more details about this payload, see the Appendix at the end of this chapter. The science payload and engineering cameras have gone through a rigorous science calibration and several calibration targets on board the rover have been used to check the calibration after arriving on Mars. All of the cameras use the same type of $1024 \times 1024$ pixel CCD. The Panoramic ($16°$ field of view) and Navigation ($45°$ field of view) Cameras on the mast view the terrain from a height of 1.5 m above the ground, compared to the Pathfinder rover's cameras, which were only 26 cm above the ground. These new rovers are able to drive as far as 40 to 140 m each day. The ability to remove the outer dusty and weathered portions of rocks and to look at rock textures at 30 times higher magnification than the rover cameras on Pathfinder, has revealed more diagnostic rock textures. The capability to identify specific minerals is also much more advanced on these rovers, and has been helpful in determining which of the rocks are volcanic.

The landing sites for the Mars Exploration Rovers were selected for their likelihood of preserving evidence for past water activity. Sedimentary rocks have the potential of answering more important science questions about Mars than volcanic rocks, because water-lain sediments might help us answer questions about whether life could have thrived on Mars. For all rocks investigated by the rovers, various hypotheses for rock origin are being considered, including a volcanic origin. If they are volcanic, then we may be able to learn about the chemistry and mineralogy of the magma and the style of eruption. Even if they all turn out to be sedimentary, the parent rock type could have been volcanic and the rocks could provide some clues about volcanism on Mars. Volcanism has shaped much of the planet's surface, so a better understanding of volcanic processes on Mars will go a long way towards revealing the geologic history of Mars and the nature of the Martian crust.

NASA is considering a discovery-driven approach to planning future Mars missions. Examples of possible pathways that could be pursued, depending on what discoveries are found by the Mars Exploration Rovers and other near-term missions, include continued global orbiter and landed exploration to characterize the diversity of Mars, landed missions in the polar regions to study climate records, a deep drilling investigation in a location where subsurface ice is expected, a long-lived long-range rover mission to explore a variety of ancient highlands rock units, or a sample return mission. The ultimate goal will be to better understand the geologic, hydrologic, and climatic evolution of Mars in order to assess habitability, as the planet evolved. For more information on the plans for future Mars missions, the reader should consult the website: `http://mars.jpl.nasa.gov/missions/`.

Indirectly, the study of volcanism on Mars can help answer questions about habitability. Although certain kinds of extreme life forms have been found living in volcanic basalt rock several kilometers deep on the Earth, this is not the most likely place we expect to find past or present life on Mars. Active volcanism brings with it a heat source, which may have helped make past Martian environments more conducive for life. One of the more likely "niches" for past life would be hydrothermal systems associated with volcanism or magma at depth. Water-deposited, fine-grained sedimentary rocks will continue to be of prime interest, because that is where fossils or other pre-biotic evidence are more likely to be preserved, but those kinds of sedimentary rocks on Mars are likely to have been originally derived from a parent volcanic rock. So, in any case, volcanic rocks are an important piece of the "Life-on-Mars?" puzzle that scientists are trying to solve.

## 7.5   CONCLUSIONS

As seen from orbit, volcanology has shaped much of the surface on Mars, although the effects are also overprinted by other processes—aeolian, impact, debris flows, volcaniclastic, fluvial, etc. Although the upcoming NASA missions are focused on learning more about water processes on Mars and possible pre-biotic or biotic activity, volcanic rocks are likely to be the original parent rock type involved. Rover missions like Pathfinder and the Mars Exploration rover missions allow us to do "robotic geology" and explore the surface. Sojourner was the first robotic mission on Mars, and from it, we discovered rocks of basaltic andesite composition, although we were not able to confirm that the rocks were volcanic. The Mars Exploration rover can drive farther and has a much more sophisticated science payload (Squyres *et al.*, 2003), and should produce a significantly better science return than Sojourner. The combination of more advanced rover and orbiter missions in the future should reveal much more to us about the volcanic history of Mars.

## 7.6   LATEST UPDATE

As of 1 June, 2004, the Mars Exploration Rovers are still in operation on Mars, after 146 Martian days and 1.8 km of driving for Spirit and 126 days and 1.2 km driving for Opportunity. Spirit has closely examined several olivine-rich volcanic basalt rocks. After solidifying as lava flows, the rocks appear to have had interaction with minor amounts of liquid water that resulted in coatings and alteration-mineral filled cracks and voids. The basaltic component is relatively unaltered, and contains the volcanic minerals olivine, pyroxene, and magnetite. In addition, Spirit has identified volcanic olivine as one of the mineral components of the soil. On the other side of Mars, Opportunity has discovered water-lain and water-soaked evaporite deposits, basaltic sands, a basaltic pebble, and a basaltic rock similar in chemistry to one of the Martian meteorites collected on Earth called ETA79001B. The science team is using the rover to study the layered rocks exposed in the walls of the crater nicknamed "Endurance", and testing whether they might be volcanic material reworked by wind or water, or more evaporite layers. Further study will be needed to understand the origin of this rock and whether it has interacted with water. So, as expected, these rovers are teaching us a lot about volcanism on Mars in addition to their ultimate goal of characterizing water-lain sedimentary rocks and the nature of past liquid water activity.

## 7.7   REFERENCES

Bandfield, J.L., Hamilton, V.E., and Christensen, P.R. (2000) A global view of martian surface compositions from MGS-TES. *Science*, **287**, 1626–1630.

Christensen, P.R., Bandfield, J.L., Hamilton, V.E., Ruff, S.W., Kieffer, H.H., Titus, T.N., Malin, M.C., Morris, R.V., Lane, M.D., Clark, R.L., *et al.* (2001) Mars Global Surveyor Thermal Emission Spectrometer experiment: Investigation description and surface science results. *J. Geophys. Res.*, **106**, 23823–23871.

Crisp, J.A., Adler, M., Matijevic, J.R., Squyres, S.W., Arvidson, R.E., and Kass, D.M. (2003) Mars Exploration Rover Mission. *J. Geophys. Res.*, **108**, 8061, doi:10.1029/2002JE002038.

Economou, T. (2001) Chemical analyses of martian soil and rocks obtained by the Pathfinder Alpha Proton X-ray spectrometer. *Radiation Phys. and Chem.*, **61**, 191–197.

Golombek, M.P., Anderson, R.C., Barnes, J.R., Bell III, J.F., Bridges, N.T., Britt, D.T., Brückner, J., Cook, R.A., Crisp, D., Crisp, J.A. *et al.* (1999) Overview of the Mars Pathfinder Mission: Launch through landing, surface operations, data sets, and science results. *J. Geophys. Res.*, **104**, 8523–8554.

Golombek, M.P., Cook, R.A., Moore, H.J., and Parker, T.J. (1997) Selection of the Mars Pathfinder landing site. *J. Geophys. Res.*, **102**, 3967–3988.

Greeley, R. and Spudis, P.D. (1981) Volcanism on Mars. *Rev. Geophys. and Space Phys.*, **19**, 13–41.

Greeley, R., Theilig, E., Guest, J.E., Carr, M.H., Masursky, H., and Cutts, J.A. (1977) Geology of Chryse Planitia. *J. Geophys. Res.*, **82**, 4093–4109.

Hodges, C.A. and Moore, H.J. (1994) Atlas of Volcanic Landforms on Mars (Professional Paper 1534). US Geological Survey, Washington, 194 pp.

Maki, J.N., Lorre, J.J., Smith, P.H., Brandt, R.D., and Steinwand, D.J. (1999) The color of Mars: Spectrophotometric measurements at the Pathfinder landing site. *J. Geophys. Res.*, **104**, 8781–8794.

McSween, H.Y., Jr., Murchie, S.L., Crisp, J.A., Bridges, N.T., Anderson, R.C., Bell III, J.F., Britt, D.T., Brückner, J., Dreibus, G., Economou, T. *et al.* (1999) Chemical, multispectral, and textural constraints on the composition and origin of rocks at the Mars Pathfinder landing site. *J. Geophys. Res.*, **104**, 8679–8715.

Nyquist, L.E., Bogard, D.D., Shih, C.-Y., Greshake, A., Stoffler, D., and Eugster, O. (2001) Ages and geologic histories of martian meteorites. *Space Sci. Rev.*, **96**, 105–164.

Rieder, R., Economou, T., Wänke, H., Turkevich, A., Crisp, J., Brückner, J., Dreibus, G., and McSween, H.Y., Jr. (1997) The chemical composition of martian soil and rocks returned by the mobile Alpha Proton X-ray Spectrometer: Preliminary results from the X-ray mode. *Science*, **278**, 1771–1774.

Rover Team (1997) The Pathfinder Microrover. *J. Geophys. Res.*, **102E**, 3989–4001.

Smith, P.H., Tomasko, M.G., Britt, D., Crowe, D.G., Reid, R., Keller, H.U., Thomas, N., Gliem, F., Rueffer, P., Sullivan, R. *et al.* (1997) The imager for Mars Pathfinder experiment. *J. Geophys. Res.*, **102**, 4003–4025.

Squyres, S.W., Arvidson, R.E., Baumgartner, E.T., Bell III, J.F., Christensen, P.R., Gorevan, S., Herkenhoff, K.E., Klingelhoefer, G., Madsen, M.B., Morris, R.V., *et al.* (2003) The Athena Mars Rover science investigation. *J. Geophys. Res.*, **108**, 8062, doi:10.1029/2003JE002121.

Wänke, H., Brückner, J., Dreibus, G., Rieder, R., and Ryabchikov, L. (2001) Chemical composition of rocks and soils at the Pathfinder site. *Space Sci. Rev.*, **96**, 317–330.

Wyatt, M.B. and McSween, H.Y. (2002) Spectral evidence for weathered basalt as an alternative to andesite in the northern lowlands of Mars. *Nature*, **417**, 263–266.

## 7.8   ACKNOWLEDGEMENTS

The writing of this chapter was carried out at JPL, California Institute of Technology, under a contract with NASA.

## 7.9    APPENDIX: MARS MISSION PAYLOAD ITEMS

Below is a list of the mission payload items mentioned in this chapter. Detailed instrument descriptions are available for Pathfinder in the papers by the Rover Team (1997), Rieder *et al.* (1997), and Smith *et al.* (1997); for TES in Christensen *et al.* (2001); and for the Mars Exploration Rover in Squyres *et al.* (2003). Initial scientific findings from the Pathfinder mission were reported in *Science* (5 December, 1997, vol. 278, pp. 1734–1774) and two special sections in *J. Geophys. Research—Planets* (25 April, 1999, vol. 104, pp. 8521–9096 and 25 January, 2000, vol. 105, pp. 1719–1865).

| Mission | Payload item | Payload characteristics |
|---|---|---|
| Pathfinder (Golombek *et al.*, 1999) | APXS on the end of a deployment mechanism on the rover. | Carries a $^{244}$Cm alpha particle and X-ray source that interacts with rocks and soils to produce backscattered alpha particles, protons, and characteristic X-rays, which are measured by detectors. Measures abundances of elements over a 5-cm diameter spot to depths of 1–100 cm. |
| | Front Stereo Rover Cameras 26 cm above the surface, mounted on the rover body under the solar panel | Monochrome (830–890 nm) imaging using a $768 \times 484$ CCD with 4 mm focal length, 127.5° cross track $\times$ 94.5° along track field of view, 2.9 mrad/pixel cross track $\times$ 3.4 mrad/pixel along track resolution. |
| | Rear Stereo Color Rover Camera 26 cm above the surface, mounted on the rover body under the solar panel | Color (500–890 nm) imaging using a $768 \times 484$ CCD with 4 mm focal length, 94.5° cross track $\times$ 127.5° along track field of view, 3.4 mrad/pixel cross track $\times$ 2.9 mrad/pixel along track resolution. 12 of every $4 \times 4$ set of pixels are "green", 2 are "red", and 2 are "infrared". |
| | Imager for Mars Pathfinder stereo panoramic lander camera 1.75 m above the surface, on the top of a mast on the lander | Spectral stereo imaging using a $512 \times 512$ CCD. Each stereo eye captures a $248 \times 256$ image with 14.4° $\times$ 14.0° field of view, 0.98 mrad/pixel, best focus at 1.3 m, depth of field 0.5 m to infinity, 23 mm focal length, $f/18$, 12 filters per eye over the 400–1100 nm range. Pointing range 360° azimuth, +90° to −67° elevation. |

| Mission | Payload item | Payload characteristics |
|---------|--------------|-------------------------|
| Mars Global Surveyor | Thermal Emission Spectrometer (TES) on an orbiter | Measures thermal emission between 200 and 1600 cm$^{-1}$, with a selectable resolution of 5 or 10 cm$^{-1}$, 2 × 3 detector array with 8.3 mrad/pixel. Pointing range 360° azimuth, +90° to −67° elevation. |
| Mars Exploration Rover (Crisp *et al.*, 2003; Squyres, 2003) | Panoramic Camera 1.54 m above the surface, on the top of a mast on the rover | Spectral stereo imaging using two 1024 × 1024 CCDs, each with a 16° field of view, 0.98 mrad/pixel, effective focal length of 38 mm, focal ratio of $f$/20, best focus at ∼3 m range with acceptable focus from infinity to within 1.5 m of the rover, filter wheel with 8 filters per eye over the 430–1000 nm range. Pointing range 360° azimuth, +90° to −90° elevation. |
| | Miniature Thermal Emission Spectrometer with a viewing port 1.4 m above the surface, on the top of a mast on the rover | Uses a single pyroelectric detector to measure thermal emission between 1,997 and 340 cm$^{-1}$, using a Fourier Transform infrared spectrometer with a resolution of 10 cm$^{-1}$ and selectable instantaneous field of view of 8 or 20 mrad. Pointing range 360° azimuth, +50° to −30° elevation. |
| | Microscopic Imager on the end of the 5 degree of freedom robotic arm. | Monochrome (400–680 nm) imaging using a 1024 × 1024 CCD with 20 mm focal length, ±3 mm depth of field at 30 nm per pixel spatial resolution covering a field of view of 3.1 cm × 3.1 cm. |
| | Mössbauer Spectrometer on the end of the 5 degree of freedom robotic arm. | Carries a vibrationally-modulated $^{57}$Co gamma-ray source that interacts with rocks/soils and results in hyperfine splitting of $^{57}$Fe nuclear levels which is characteristic of Fe bonds and Fe oxidation states within specific minerals. $\gamma$ and X-ray detectors measure backscattered radiation at energies of 6.4 keV and 14.4 keV while the $^{57}$Co source velocity is changing. Measures abundances of elements over a 1.5 cm diameter spot to depths of 200–300 μm. |

*continued*

| Mission | Payload item | Payload characteristics |
| --- | --- | --- |
| | APXS on the end of the 5 degree of freedom robotic arm. | Carries a $^{244}$Cm alpha particle source that interacts with rocks/soils to produce backscattered alpha particles and characteristic X-rays, which are measured by detectors. Measures abundances of elements over a 4 cm diameter spot to depths of 1–100 μm. Less $CO_2$ background in X-ray mode than the Pathfinder APXS and much higher signal-to-noise for the same integration time. |
| | Rock Abrasion Tool on the end of the 5 degree of freedom robotic arm. | Tool with spinning diamond-grit grinding teeth that abrades the outer 5 mm of rock surfaces over an area 4.5 cm in diameter. |
| | Stereo Navigation Cameras 1.54 m above the surface, on the top of a mast on the rover. | Acquires monochrome (centered on 650 nm, 140 nm FWHM) stereo $1024 \times 1024$ pixel images using two CCDs with 14.7 mm focal length, $f/12$ optics, 45° field of view, 0.82 mrad resolution, depth of field 0.5 m to infinity, best focus at 1 m. |
| | Front and Rear Stereo Hazard Cameras mounted on the rover body, under the rover solar panels. | Acquires monochrome (centered on 650 nm, 140 nm FWHM) stereo $1024 \times 1024$ images using two CCDs on the front and two on the back of the rover, each with 5.6 mm focal length, $f/15$ optics, 124° field of view, 2.1 mrad/pixel resolution. |

# 8

# Io, a world of great volcanoes

***Rosaly Lopes*** (Jet Propulsion Laboratory, California Institute of Technology)

*I was born and raised near the famous Ipanema Beach in Rio de Janeiro, Brazil. Ever since I can remember, I wanted to become an astronomer and work for NASA. I probably got my inspiration from the race to the Moon in the 1960s. I thought that exploring space was the most exciting thing anyone could do, and I wanted to be part of it. My teachers and relatives thought it was very impractical to study astronomy, but I stuck with my ambition and left Brazil for University College London, from where I graduated in astronomy in 1978. I had become very interested in planetary geology by that time and decided to pursue that field for my PhD at the University of London Observatory. I studied volcanoes on Earth and Mars for my thesis research and got to travel to Mount Etna and Hawaii. I simply fell in love with volcanoes and since then I have traveled to many of the Earth's active volcanoes. I came to work at NASA's Jet Propulsion Laboratory in 1989, initially as a postdoctoral researcher. I joined the Galileo Mission in 1991, and led the planning and data analysis for Io using Galileo's Near-Infrared Mapping Spectrometer instrument. I had the thrill of discovering 71 previously unknown volcanoes and of being a co-discoverer of the extremely high-temperature magmas that may be unique to Io. I now work on the Cassini Mission to Saturn and still continue with research on Io's volcanoes, analyzing the rich data set returned by Galileo.*

## 8.1 ACTIVE VOLCANOES!

Io (Figure 8.1; see color section) came into my life on a gray English afternoon in March, 1979, when news of anything hot—even in another world—was good news. "They have discovered active volcanoes on Io!" I looked up from my desk at my thesis advisor, John Guest, who had just got off the phone with an American colleague. I distinctly remembered trying to change mental gears from images of the great but dead Olympus Mons volcano on Mars, which I was mapping. I was a first year graduate student—very green as a research scientist and only vaguely aware of the Voyager 1 spacecraft's fly-by of the Jupiter system. We worked at the University of London Observatory, during what my son thinks of as the Dark Ages (pre-internet), before web sites showed us what was happening around the world. The Voyager 1 fly-by had not been, until that day, a front-page event in England.

As my brain frantically tried to work out how this moon of Jupiter could have active volcanoes, John rushed off to the fax machine and came back with a smudgy image of Io's volcano Ra Patera (Figure 8.2). My first glimpse at Io's surface was, therefore, somewhat unimpressive. I could see a dark caldera-like ink blob, and smudges that could be lava flows, but still things didn't make sense. How did "they"—the Voyager scientists—know that this volcano was active? Besides, what I had learnt in planetary geology classes went contrary to active volcanism on a moon roughly the size of ours. Io, like our moon, should have cooled a long time ago and be a dead world by now.

It turns out that, to this day, nobody knows if Ra Patera was erupting during the Voyager fly-by. And if I had been assiduously reading the latest journals (though I cannot be sure that the relevant issue of *Science* had even arrived at our library), I would not have been so surprised at the news. The story of the discovery of Io's volcanoes is truly amazing and, many years later, I heard parts of it directly from some of the leading players.

It starts back in the early 1970s, when astronomers observing Io using telescopes discovered some peculiar facts about this moon. Bob Brown from Harvard University discovered a cloud of sodium gas surrounding Io (Brown, 1976), which was later found to have sulfur and oxygen as well. Observations made by the Dutch astronomer W. Wamsteker (Wamsteker *et al.*, 1974) showed that the spectrum of Io closely resembled that of sulfur and they suggested that sulfur and sulfur compounds might be abundant on the surface. What was particularly surprising was the fact that Io's spectrum lacked water ice, unlike those of the other Galilean satellites. Something had boiled off the water on Io and now there was sulfur instead. Io was different, but why? In 1978, Bob Nelson and Bruce Hapke (Nelson and Hapke, 1978) from the University of Pittsburgh even went as far as to suggest that sulfur compounds might be produced on Io, "in the vicinity of a volcanic fumarole or hot spring". This suggestion was met with great skepticism within the science community.

Just prior to the Voyager encounters in 1979, another science result presaged the discovery of active volcanism. Fred Witteborn and colleagues (Witteborn *et al.*, 1979) reported a telescopic observation of an intense temporary brightening of Io in the infrared wavelengths from 2–5 μm. They explained it, although with some skepticism, as a thermal emission caused by part of Io's surface being at a temperature of about 600 K, much hotter than the expected daytime surface temperatures of about 130 K. How could part of Io's surface suddenly become hotter than its surroundings?

The idea of active volcanism was hard for scientists to accept, because the only place where we had seen active volcanoes was on the Earth. Thus, we all thought that little worlds the size of Io must be dead like our own Moon. The answer to Io's

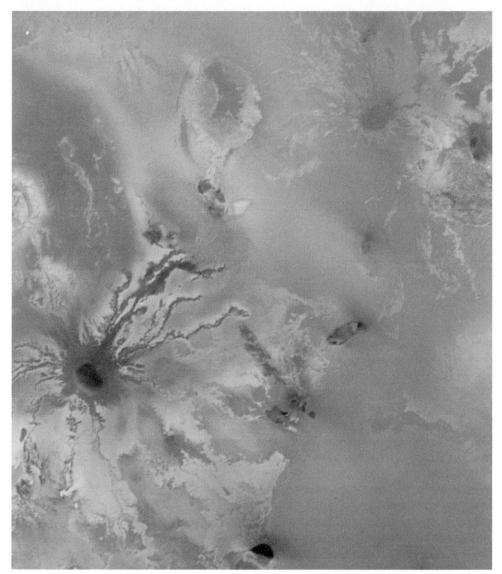

**Figure 8.2.** My first view of Io's surface was a smudgy version of this image of Ra Patera, taken by the Voyager 1 spacecraft in March 1979. The image shows a dark caldera in the lower left, surrounded by flows which were thought to be sulfur. The image is about 1,000 km wide.

mystery came in the form of an impeccably-timed research paper in *Science* by Stan Peale, Pat Cassen, and Ray Reynolds (Peale *et al.*, 1979), published only 2 weeks before the frenzied Voyager encounter. This remarkable work predicted massive heating of Io's interior due to distortions in the moon's shape by the massive gravity of Jupiter, coupled with the gravity of the other Galilean satellites that, although smaller, are closer to Io. In fact, Io finds itself in a veritable "tug-of-war" between Jupiter and the moons Europa, Ganymede, and Callisto. Stan Peale and his colleague wrote that, "widespread and recurrent surface volcanism" might occur on Io. It was a bold statement, but it hit the bullseye.

The prediction had been made, but there is nothing as effective as a picture to make it reality. The two Voyager spacecraft were approaching Jupiter, and Voyager 1 flew by Io in early March. When the first Io images were transmitted down to Earth and arrived at JPL, members of the Voyager Imaging Team, led by Brad Smith, saw a colorful, strange surface that they thought looked like a pepperoni pizza. Io thus earned the nickname "pizza moon", which has stuck to this day. The dark, olive-like features on the pizza looked like volcanic calderas. Most surprising was what they *didn't* see: not a single impact crater could be found in the images. Since the density and sizes of impact craters correlate well with surface age for solid surfaces of the solar system (see Chapter 5), this meant that Io's surface was very young. Some process had caused the craters to be erased (on Earth, erosion and plate tectonics played a major role). Since there seemed to be volcanic craters on Io's surface, the conclusion was that volcanism had been active in the not too distant past. In geological terms, "not too distant" can mean thousands or even hundreds of thousands of years.

It would be another four days before Io revealed itself to have *live* volcanoes spewing debris into outer space while Voyager flew by. A young navigation engineer at JPL, Linda Morabito (Morabito *et al.*, 1979), noticed a peculiar, umbrella-shaped mark on Io's limb as she analyzed images used for optical navigation (Figure 8.3). At first, she thought there was another satellite behind Io, but after checking and rechecking the data with her team leader, Steve Synnott, she became convinced that something else was happening, possibly some kind of cloud coming from Io. The location of the cloud seemed to be a heart-shaped red feature on the surface, possibly volcanic in origin. It was, however, hard to imagine that a volcanic cloud could reach some 270 km (170 miles) above the surface. The confirmation soon came. Computer enhancement of other images showed more "clouds" or plumes (the image obtained by Linda was deliberately overexposed, and this was why the plume showed up). Independently from the navigation and imaging teams, scientists working with Voyager's infrared instrument, IRIS, found enhanced thermal emission from parts of Io's surface, showing that a few places appeared to be much hotter than the background surface. When the most prominent of these "hot spots" was found to coincide with one of the erupting plumes, there was no doubt left at all. Io had active volcanoes!

What materials were the volcanoes erupting? The IRIS instrument detected sulfur dioxide from the plume over the volcano now called Loki (see Figure 8.4). Back on Earth, Bob Nelson, now working at JPL with Bill Smythe, took a new look at Io's infrared spectrum. Could the sulfur dioxide be condensing from the plumes and be mantling Io with sulfur dioxide frost? Back in his lab, Bill obtained a spectrum of $SO_2$. There was a match, quickly reported in less than one page in *Nature* (Smythe *et al.*, 1979). In the same issue of this scientific magazine, Carl Sagan (Sagan, 1979) proposed that Ra Patera—which I remembered from the fax

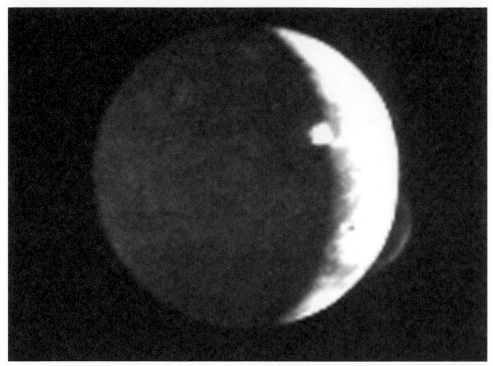

**Figure 8.3.** Active volcanism on Io was discovered from this image, taken by Voyager for navigation purposes on 8 March, 1979, when the spacecraft was about 4.5 million kilometers from Io. The image shows two simultaneously occurring volcanic eruption plumes. One can be seen on the limb (lower right), rising more than 260 km above the satellite's surface. The second can be seen on the terminator (shadow between day and night) where the volcanic cloud is catching the rays of the rising sun. The dark hemisphere of Io is made visible by light reflected from Jupiter.

machine—was a volcano spewing out sulfur flows. The vivid colors of Io's surface—yellows, oranges, reds, and browns—were interpreted as those of sulfur cooled rapidly from different temperatures. Sulfur seemed to be everywhere and it was a logical conclusion that the volcanoes on Io were sulfur volcanoes.

Despite the mounting evidence for sulfur volcanism, some scientists disagreed and proposed that sulfur formed only a thin coat over the surface and that the erupting lavas were silicates, probably basalts, similar to those found at Kilauea in Hawaii and many other terrestrial volcanoes (see Chapter 2). This debate—sulfur or silicates—would go on for years.

When the second Voyager spacecraft flew by Io four months after Voyager 1, Io showed some significant changes. The Pele volcano, whose plume was the first to be spotted, showed a large change in its red heart-shaped plume deposit, the indentation had been filled in and the deposit was now oval. The total changed volcanic area was more than 10,000 km$^2$—about the size of the island of Corsica (France) or half the

**Figure 8.4.** Voyager 1 image of Io (March 1979) showing the active plume of Loki on the limb. The heart-shaped feature south-east of Loki consists of fallout deposits from the active plume Pele. The Pele deposit had filled out and had become oval shaped by the time Voyager 2 flew by Io four months later.
See cover for color version.

size of New Jersey, USA. The active plume over Pele was no longer active, but two new ones had started elsewhere. The two brief glimpses by these spacecraft presaged what Io would reveal to the Galileo spacecraft: an ever-changing surface that is a paradise for volcanologists but surely a nightmare for cartographers.

Back in England, I read about Io with great interest, but stuck to volcanoes on Mars and Earth for my PhD. However, the smudgy Io image from the fax machine had awakened something in me that was more than scientific curiosity. I daydreamed about what it would be like to work on a flight project like Voyager, to be part of a team that saw images from other worlds before anybody else. Io had set my course. A decade later I left England for the U.S.A., to continue my research into planetary volcanoes at NASA's JPL. I had the good fortune to befriend Adriana Ocampo, a member of the Near-Infrared Mapping Spectrometer (NIMS) team on the Galileo spacecraft. Galileo was launched in 1989 to investigate Jupiter and its moons, to find out answers to the many questions posed by Voyager. Through Adriana, I found out that the NIMS team needed help with the planning and design of observations. I was lucky enough to be hired and to have special responsibility for Io. Galileo turned out to be an amazing mission both for its scientific return and its endurance. Best of all, being part of a flight project team turned out to be much more exciting than I had envisioned in my student days.

## 8.2  GALILEO GOES TO JUPITER

Galileo was the first spacecraft to orbit Jupiter and to make repeated observations of its moons. A big, complex spacecraft, Galileo had a plethora of scientific objectives, including investigations of Jupiter's atmosphere, its huge magnetic field, and its moons. Io was a small but important part of this investigation. For the remote sensing instruments, such as the camera (called the Solid State Imaging System, SSI), NIMS, and the Photopolarimeter Radiometer (PPR), Io was one of the top priorities. Each instrument had its own team of scientists, and negotiations about observing time and data that could be returned were often contentious. Galileo's main communications antenna never fully deployed, so our data trickled down painfully slow. There were a lot of limitations on what we could do and it was up to us scientists to come up with a very limited number of observations that could answer many science questions. Three of us—Alfred McEwen for SSI, John Spencer for PPR, and myself for NIMS—did the planning for Io observations for our respective instruments, keeping in mind that cooperative observations would yield the best science. The camera (SSI) could observe in the visible and get to the near-infrared using a 1-$\mu$m filter, NIMS had a wavelength range from 0.7–5.2 $\mu$m but lower spatial resolution (50 times lower than the camera's), and PPR could observe at even longer wavelengths than NIMS in the infrared. To study volcanism, the combination of wavelengths was very desirable. PPR could measure cooler temperatures than NIMS, which could not detect temperatures below 180 K. SSI could image the surface and reveal the geomorphology and surface color and, when observing Io in eclipse, measure temperatures hotter than 700 K. NIMS could detect absorptions in its wavelength range that included $SO_2$ and other compounds. The instruments complemented each other well and thus we began years of planning, followed by spectacular results that just kept on coming.

What we knew about Io from Voyager and later ground-based observations determined what type of observations we would make and what we hoped to find out. A major question was the composition of Io's volcanism, as the silicate versus sulfur debate raged on. Voyager observations favored sulfur volcanism, because of surface colors, the detection of $SO_2$ from a volcanic plume, and particularly because no temperatures higher than about 600 K were detected by the infrared instrument. In fact, temperatures seemed to be around 400 K. Temperature of the molten material is a tell-tale sign of composition. Basalts begin to melt at temperatures of around 1,300 K, but sulfur melts at about 400 K. If Io's volcanoes were spewing out basalts, as some scientists thought, why did Voyager not detect higher temperatures?

There were, however, some good arguments on the silicate side. Io's density, similar to that of the Earth's moon, was consistent with a silicate composition. The topography of Io, and the morphology and topography of the volcanic features such as calderas, were similar to those found on basaltic volcanoes on Earth. The IRIS instrument on board Voyager was not designed to be sensitive to high temperatures. The temperature argument was weak, because silicate lavas such as basalt cool very quickly once they are erupted. This is why it is possible to walk over moving lava here on Earth, as long as the cool crust is strong enough to support one's weight (and one's boots do not melt). The fact is, we could not tell if these lavas on Io were molten sulfur or basalts that had cool crusts.

As Galileo was getting ready to be launched, astronomers continued to observe Io from the ground, now using infrared cameras. A year before the launch, Torrence Johnson, the Galileo project scientist, and colleagues published a paper (Johnson *et al.*, 1988) reporting temperatures from an eruption on Io around 1,000 K—far too high for sulfur. As Galileo got on its way in 1989, we knew that sulfur and silicates most likely played important roles in Io's volcanic activity and we knew of at least a dozen hot spots (volcanoes from which enhanced thermal emission was taking place) which had been observed by Voyager or from ground-based telescopes.

After a long and tortuous journey, Galileo arrived at Jupiter in December 1995. The spacecraft flew within 1,000 km of Io's surface, the only close fly-by of Io that had been planned for the mission. It turned out to be a very difficult day for us in the remote sensing teams, particularly those of us whose main investigation was to be Io. Because of a problem with the on-board tape recorder, no remote sensing observations of Io were recorded or transmitted to Earth. That was, we thought, our only chance to get close to Io—and we had lost it.

## 8.3   HOT LAVA THROUGH GALILEO'S EYES

Galileo's planned orbits around Jupiter kept it away from Io for the duration of our prime mission, which lasted until December 1997. During these first two years, the spacecraft stayed at distances of over 200,000 km of Io, flying close to the surfaces of the other Galilean satellites. The main reason for staying away was the extremely harsh radiation environment near Io created by Jupiter's powerful magnetic field, which would bombard the spacecraft with about 4 Megarads in a single day. This is about 4,000 times the lethal dose for a human, and not healthy for spacecraft either.

Thus, the phase of monitoring Io's eruptions from great distances began and, thanks to a NASA extension of the mission's duration, lasted until October 1999. Every few months, when the spacecraft came closest to Io during each orbit around Jupiter, we made observations with the remote sensing instruments (Figure 8.5; see color section). Our first observations happened in June 1996. The camera saw significant surface changes since the Voyager days but, surprisingly, Io's most powerful volcano, Loki, had hardly changed in its appearance. We had been curious to see Carl Sagan's "sulfur flows" at Ra Patera with Galileo's camera, but they were gone, as a new eruption had covered them all. Our first NIMS observation proved to be very exciting. Before the data came, a colleague of mine had asked me if I thought we would find any new hot spots (active volcanoes) on Io. I replied that I hoped we would. On that first observation alone there were 13 new hot spots. During the next four years we found many more and this turned out to be the most fun part of the mission for me. As the person who first analyzed all the NIMS Io observations, I detected and kept track of the ever-increasing number of active volcanoes (Lopes–Gautier et al., 1999). Comparison of the NIMS thermal data with the camera's higher resolution images confirmed a correlation observed from Voyager data: hot spots corresponded to dark areas on the surface. Basalts are dark and thus the evidence for silicate eruptions kept getting stronger, but color alone does not give definite answers, as some forms of sulfur can be black. However, NIMS data, as well as images of Io in eclipse taken by the camera, showed numerous hot spots where the temperatures were too high for molten sulfur—over 1,000 K. The consensus early on among the Galileo scientists was that Io's eruptions were predominantly of basaltic lavas. However, Io had a major surprise in store for us.

Astronomers already knew that some of Io's eruptions were much more powerful than others. "Outbursts" had been observed from ground-based telescopes (Spencer et al., 1997). These are events that are so powerful that they double the total flux from Io at a wavelength of 5 μm. We knew from surface changes that some eruptions had left large plume deposits on Io. By mid-1997, we had not yet caught a major eruption in progress. NIMS had, however, been observing what seemed to be an ordinary hot spot located in the eastern part of the Pele red plume deposit. This hot spot, named Pillan, appeared inconspicuous in an image taken by the camera in April 1997, but when the camera looked at the area again in September, an astonishing change had occurred (Figure 8.6; see color section). Pele's red ring deposit now showed a "black eye" some 400 km across, centered on the Pillan hot spot. Our inconspicuous volcano had erupted spectacularly, depositing dark materials (probably ash and lava fragments) in an area large enough to cover the state of Arizona. But the biggest surprise came when we analyzed observations of Pillan taken in June of that year. By sheer luck, NIMS and SSI had looked at Io while the Pillan eruption was raging.

NIMS and SSI have complementary wavelength coverage, so we knew that it was useful to have the two instruments observing one after the other so that our observations were taken essentially at the same time and from the same look angle.

Therefore, we had planned pairs of Io observations whenever possible. During Galileo's June 1997 Io campaign, SSI planned to observe Io while it was eclipsed by Jupiter, so that hot spots could be detected using the camera's filter near 1 μm. I had not planned to follow this observation with one by NIMS, because the eclipse happened when Galileo was so far away from Io (over 1 million km) that the NIMS observation would be very low in spatial resolution. Therefore, I planned my observations when the spacecraft was closer to Io. But serendipity intervened, and competition for the time closer in forced me to give up an observation. I moved it so that it happened during the eclipse time, without many hopes that the results would be interesting. But space missions are full of surprises and this low-resolution observation turned out to be one of the most valuable we have, as it resulted in the most interesting discovery about Io's volcanism made by Galileo. We combined the camera's 1-μm images with the NIMS wavelength range, constructed a spectrum from 1–5.2 μm, applied a black body radiation model and found that the Pillan eruption had a temperature of at least 1,825 K (McEwen *et al.*, 1998). This was hotter than anything we had ever observed on Io and, more surprisingly, hotter than anything ever observed on Earth. Not only are temperatures in this range too high for the material to be sulfur, they are too high for ordinary silicates like basalt.

What are these lavas? We still cannot tell for sure, but the most logical explanation is that they are similar to terrestrial lavas called ultramafic, which erupted on Earth in pre-historic times, mostly billions of years ago. These lavas are rich in magnesium and melt at temperatures in the range 1,700–1,900 K. If Io's lavas are indeed of the same type (Williams *et al.*, 2000), we can learn important things about how ultramafic eruptions happened on Earth by observing them happening on Io. It is like having a window into the Earth's distant volcanic past.

How common are these lavas on Io? It is possible that all of Io's volcanoes erupt these very hot lavas but, unfortunately, such temperatures turned out very difficult to detect with confidence. The first problem is that lava cools very quickly once it erupts, so the instruments need to be looking at just the right time, when the eruption is vigorous and the cooled crust is being broken up enough to expose the red-hot, molten material. If the exposed areas are very small, and the spacecraft is observing from far away, the instruments will not be able to detect them. Another problem is that Galileo's instruments were not really designed to measure such high temperatures and, as we found out later during the close fly-bys, a vigorous eruption can easily saturate the instruments' pixels. Measuring these ultramafic-range temperatures with confidence turned out to be a hard job and Pillan is still the only volcano where we can be sure that these very hot lavas erupted.

Temperature can give a constraint on composition, but without other data, we cannot say for sure that Io's lavas are ultramafic. We knew that the best way to determine composition would be to obtain spectral data of the dark regions, where the freshest lavas are. Unfortunately, the great distances of the spacecraft from Io during the early part of the mission did not allow NIMS to resolve any of the dark areas because of their small size. The camera's spatial resolution was much higher and they could resolve the dark patches in different filters. Paul Geissler and his colleagues found a spectral absorption at about 0.9 μm, which they argued indicated lavas having magnesium-rich orthopyroxene (Geissler *et al.*, 1999). This is good support for an ultramafic composition. We looked forward to what NIMS observations taken during the fly-bys, close enough to resolve the dark patches of lava, would reveal. As a spectrometer, NIMS was better equipped to detect the tell-tale signs of different minerals. The low-resolution observations had not shown any new compounds. Io's surface is covered by $SO_2$ frost and it turned out to be impossible to find large patches that did not have the frost masking whatever else might be underneath. We clearly needed to get closer to Io and to take a close look at the relatively small areas of active lavas to find out what they were made of.

As we neared the end of the Galileo Europa Mission (as the extended mission was named), we prepared to make two risky close fly-bys of Io. We knew the spacecraft and instruments might be severely damaged—or even die—if we flew inside the high-radiation environment close to Io, as the engineers had not designed it to withstand more radiation than the dose we encountered during the first fly-by back in 1995. But since it was the end of the extended mission, even the traditionally conservative engineers were willing to take a chance—or two. Thus two close Io fly-bys were planned for October and November 1999. Then things took a great turn for the better when NASA approved another year's extension for Galileo, until the end of 2000. The main reason for the extension was that the Cassini spacecraft, on its way to Saturn, would fly by Jupiter in December 2000. It would be highly valuable to have two spacecraft taking simultaneous measurements of the Jupiter environment—if Galileo survived until then.

We planned another Io fly-by for February 2000 (which would be Galileo's fourth), even though our project manager at the time, Jim Erickson, told us we were dreaming. The spacecraft would not survive such a high accumulation of radiation dose. But he also agreed we should risk it. It would require luck, he said, "but we have had a lot of luck so far". He was right. Galileo not only survived the 1999 and 2000 fly-bys, but was able to come back for two more successful Io fly-bys in 2001. The new phase of the mission—which was named after my suggestion "Galileo Millenium Mission"—was successful beyond all our expectations.

## 8.4  VOLCANOES CLOSE-UP

Our first chance to take images and spectra of Io close-up happened on 11 October, 1999. Because of the loss of Galileo's high-gain antenna early in the mission, our data had to be stored on board the spacecraft and sent to Earth painfully slowly (our maximum data rate was 160 bits per second), using the low-gain antenna. I could not wait to see the NIMS data but, unfortunately, the camera's images were scheduled to be sent back first. The start of the "playback" time coincided with a meeting of the American Astronomical Society's Division for Planetary Sciences, that year held in the quaint Italian town of Abano Terme. My colleague Alfred McEwen stayed home in Tucson waiting for his first close-up images of Io. I looked at the schedule and realized that I could go to the meeting and be back before any NIMS data came down. But missions always have surprises. Some of the first camera images sent to Earth were horribly scrambled. The culprit was the feared radiation dose, which affected one of the camera's data collecting modes. Because of this, the playback schedule was changed and during the late afternoon, local Italian time, on 14 October I found out that the first NIMS observation, targeting the Loki volcano, had come down. I was in Italy without a computer and the observation was in our system at JPL. I just had to see it somehow. Thus begun a rather amusing evening, in which I rounded up three of my NIMS colleagues—Bill Smythe, Lucas Kamp, and Frank Leader—during the conference's banquet. Bill had a laptop computer and Lucas and Frank could do the required processing of the data. We went back to my hotel around midnight to try to get the data. When I asked the elderly hotel clerk for my room key, I got a really dirty look and wondered if I should try to explain why I had three men with me. Luckily we got past him and nobody from the hotel found out how we unscrewed the old rotary-dial phone from the wall and rigged our internet connection. Around 3 am we saw Loki's infrared spectra—at high spatial resolution for the first time—and measured a lava temperature of about 300 K. Not very hot, but still exciting, most of all because it showed us that NIMS was still alive and getting great data.

It would be a couple of days before we found out that NIMS had indeed suffered some irreversible damage from the radiation. The instrument had a moving (scanning) grating that, together with 17 detectors, allowed us to collect up to 408 wavelengths in the range 0.7–5.2 μm (Carlson *et al.*, 1992). Before the fly-bys we had already suffered the degradation of 2 detectors and the loss of 3 others. But now the grating was stuck and we were limited to only 12 wavelengths in the range 1–4.7 μm (Lopes–Gautier *et al.*, 2000). This was still good enough to measure temperatures, because we can fit a black body to these few points, but our hopes of determining the composition of the lavas using spectroscopy were dashed. What turned out to be astonishing was how much science we could still accomplish with a crippled instrument. Meanwhile, the camera team had found out how to unscramble their images to the point that they provided useful information, and another camera mode, which worked without problems, was used for the subsequent fly-bys. Galileo came back close to Io again in November 1999, February 2000, August and October 2001, acquiring spectacular data. Unfortunately, our last fly-by in January 2002 did not work well, as the spacecraft went into "safing" mode just prior to our Io observations. "Safing" happens when the spacecraft senses that something is wrong and shuts off its observing sequence to "call home". These events were not unexpected during the high-risk fly-bys and one had caused us to lose the closest approach observations of Io in November 1999. On that occasion, the spacecraft operations team members were called from their Thanksgiving dinners and were able to restart all systems and send up a truncated science sequence in time to get some of the observations. Unfortunately, in January 2002, the safing happened at the worse possible time and we could not recover fast enough.

Despite all the problems, the data set gathered from these close fly-bys was rich and full of surprises. Io was much more volcanically active than we had realized from the distant observations. The closer we got, the more hot spots we could detect. I still keep a master list of hot spots on Io, in which I incorporate all known hot spots including the many found by myself with NIMS, and those found by my colleagues on other teams. As of the end of 2003, we know of at least 166 hot spots on Io. A reporter recently asked me how many I had personally detected. Some of the detections were made nearly simultaneously by myself and by my colleagues on the camera team, so it is hard to tell (and not really important) who found a new hot spot first, but I estimate that I detected 71 of them. Perhaps I can claim to have found more active volcanoes than anybody else so far, though my colleagues studying undersea volcanism on Earth may disagree (see Chapter 3). Detecting new hot spots on Io may not constitute a truly major scientific finding, but it was certainly fun.

One of the most exciting detections was of a very bright, powerful new hot spot during the August 2001 fly-by. This was a particularly interesting fly-by because a plume coming out of a volcano called Tvashtar had been detected the previous December by the Cassini spacecraft, when it flew close to Jupiter on its way to Saturn. If this 385 km high plume was still active, Galileo's August fly-by would take the spacecraft right through it. The question was, would it still be active and if so, would it damage Galileo?

Io's plumes seem to come in two main types, intermittent or long-lived. The intermittent plumes, of which Pele and Tvashtar are examples, reach great heights but turn on and off. Long-lived plumes are smaller in height. We worried whether the spacecraft could survive the fly-by if the Tvashtar plume was erupting at the time. After a lot of discussion, we decided to take a risk and not change the trajectory. We based our decision mainly on the likelihood of low particle density in the plume and the fact that there was a reasonable chance that it might not be active at all. When the fly-by took place, taking the spacecraft close to Io's north pole on 6 August, 2001, Galileo came through just fine.

When the camera team analyzed their images, they discovered a huge plume— some 500 km high, the largest ever seen on Io—but its location did not match Tvashtar's. As camera team members tried to pinpoint the source of the plume, I analyzed a new observation from NIMS. The infrared image coved a large region of Io, pole to pole. It showed a new, very bright hot spot located about 600 km to the south-west of Tvashtar (Figure 8.7). This was clearly where the huge plume was coming from. The new volcano, which is now known as Thor, was in the midst of a gigantic eruption. We had worried that the Tvashtar plume might be active (and the data indicate that it was), but Io had thrown a new, larger plume in our path. In fact, Galileo was, for the first time, able to catch a "whiff" of a plume. The plasma science experiment, led by my colleague Louis Frank (Frank and Paterson, 2001), detected $SO_2$ molecules within minutes of their escaping from the plume vent. Commenting on the results, Frank said that Galileo had smelled the volcano's hot breath and survived.

The Thor eruption is an example of the very violent eruptions that Io can produce. After analyzing many visible and infrared images, we now classify the eruptions on Io into three major types, named Pillanian, Promethean, and Lokian.

**Figure 8.7.** Volcanic hot spots, including a bright one never seen before, are seen all over an infrared image (left) of Jupiter's moon Io, taken by Galileo NIMS on 6 August, 2001. The bright, new hot spot (Thor, indicated by the arrow) was the source of a towering volcanic plume. This infrared image shows the brightness of features at the wavelength of 4.4 μm, which detects heat from Io's many volcanic eruptions. An earlier image from Galileo's camera (right) is shown for context. Many volcanic hot spots appear in the infrared image as bright regions. This eruption of Thor was so vigorous at the time of the observation that some pixels (shown in black) were saturated.

Thor and Pillan are typical of the most violent eruptions, which tend to be short-lived, at least during their most intensive phase. The Tvashtar volcano, which Galileo caught erupting in November 1999, is another example. That eruption was so intense that it saturated both the camera and NIMS images, but we managed to reconstruct what was happening from the distribution of the camera's bleeding pixels. A fissure had opened in the caldera, from which a curtain of hot lava spewed out. Some months later, the tall plume that worried us so much was erupted. Pillanian eruptions tend to produce tall but generally short-lived plumes.

Promethean eruptions are far less violent. The typical example is the Prometheus volcano (Figure 8.8), also called the "Old Faithful" of Io, because its plume appears to be constantly erupting. The plumes produced by this type of eruption are long-lived but small by Io's standards, meaning less than 200 km in height. These eruptions have long lava flows (e.g., the Prometheus flow is nearly 100 km long). The plumes appear to be caused by the interaction of lava flows with frozen underlying $SO_2$, as proposed by my colleague Susan Kieffer (Kieffer *et al.*, 2000).

Lokian eruptions are those confined inside calderas and they are by far the most common eruption type on Io. They are named after Loki, Io's most powerful volcano and largest caldera (Figure 8.9; see color section). One of the surprises from Galileo's fly-by data was that many of these calderas appear to be lava lakes. Lava lakes form on some volcanoes on Earth, such as Kilauea in Hawaii

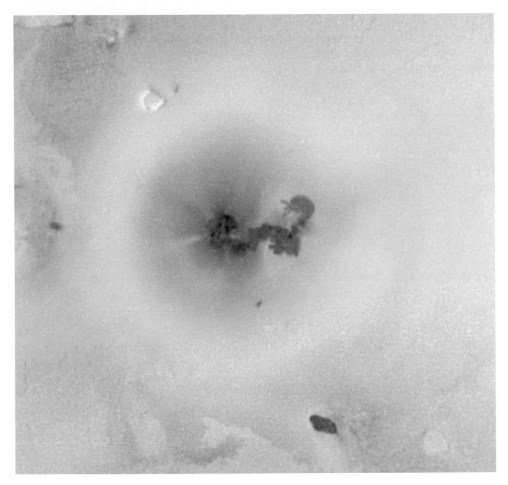

**Figure 8.8.** Prometheus could be called the "Old Faithful" of the outer solar system, because its volcanic plume has been visible every time it has been observed since 1979. The long-lived plume has produced a ring-like deposit of bright white and yellow material that is rich in $SO_2$ frost. The plume was remarkably stable in its height and shape, but between the Voyager fly-bys and the time of Galileo's arrival at Jupiter, its source shifted about 70 km (44 miles) to the west. This type of plume eruption is thought to be a result of lava advancing over $SO_2$ frost. (Image is about 600 km across.)

and Erta Ale in Ethiopia. They are often long-lived, lasting years or decades, like many of Io's volcanoes seem to be. Galileo observations showed a characteristic of lava lakes in its thermal images—the presence of "hot edges" near the caldera walls. This happens because the lava forms a cool crust over the lake, but because there is some movement of the underlying molten material, the crust tends to "slosh around" and break up as it hits the caldera walls, much like the foam on bathtub water. The most spectacular signature of these "hot edges" was found in a NIMS image of Loki taken during Galileo's last successful fly-by.

I planned this NIMS observation of Loki in part to test a model of Loki's behavior that had been proposed earlier that year by astronomer Julie Rathbun and colleagues (Rathbun *et al.*, 2002). They concluded that Loki is a lava lake that overturns periodically. In their model, the crust of the lava lake founders at the western edge and causes a wave of fresh lava to flood the caldera, traveling east. The periodicity of the event is quite interesting. It could be studied because Loki is usually bright enough to be observed from Earth. Julie analyzed years of ground-based observations and found that the brightenings that happened at Loki (related to the flooding events), which cause its flux at 3.5 µm to increase by a factor of 10 or more, happen about every 540 days.

If Julie and her colleagues were right, the caldera should show a thermal pattern of lava moving from west to east. The NIMS observation supported the model very well. It showed a "hot edge" along the western caldera wall, where they predicted the crust would founder. The thermal map from NIMS data also showed that the oldest (coolest) crust was next to the hot edge, getting progressively hotter (younger) towards the east. This supports the idea of lava moving (and cooling) from west to east. The outer edges of the lava lake, close to the walls and the central island, are hotter than the middle, which again agrees with a lava lake model, as we would expect abrasion against the caldera walls to break up the crust. We think that Loki may be a gigantic lava lake, but there are still aspects of its eruptions that remain puzzling. One is the absence of very high temperatures, even at the hot western edge (Figure 8.9; see color section). Many other volcanoes on Io show hotter temperatures and, given the frequency of observations of Loki, it is somewhat surprising that higher temperatures have not been detected. Another puzzle is the reason for a central "island". Loki and some other Io calderas, such as Tupan (Figure 8.10; see color section), show a central "island" which at least in the case of Loki and Tupan, appears to be cold (Lopes *et al.*, 2004). We assume that the islands are topographically higher than the lava lakes, otherwise they would have been covered by lavas. We also think that the islands are anchored and not floating, otherwise we would have expected changes between the images Voyager took in 1979 and the Galileo images nearly two decades later. Terrestrial calderas do not show similar features and the origin of these "islands" remains mysterious. I am turning my attention again to volcanoes on Earth, trying to find clues here about caldera formation that may help us better understand Ionian calderas.

## 8.5   WHAT NEXT?

Io is undoubtedly one of the strangest planetary bodies we know. We could learn a lot more from a new spacecraft mission, but sending a mission to Io is not easy because of the intense radiation environment. Although there are currently no plans for a dedicated Io mission, two opportunities to observe Io from distances similar to those attained by Galileo before the fly-bys are in the works. One is the Pluto-bound New Horizons, due to be launched in 2006. Current plans call for the spacecraft to obtain a gravity assist from Jupiter on its way to Pluto, making observations of Io and the other moons on its way. Another opportunity will perhaps come in the next decade, when the Jupiter Icy Moons orbiter launch is planned. This mission will orbit Callisto, Ganymede, and Europa, and obtain Io observations from the orbits of these other moons. Meanwhile, techniques for observing Io from Earth continue to improve and these ground-based observations will serve as our eyes on Io's volcanoes for years to come. As for me, I continue to analyze the Galileo data, but as a member of the Cassini Mission flight team, I am now turning my attention towards the moons of Saturn, where cryovolcanism might be taking place (see Chapter 9). There are new, strange worlds to explore and I feel that we have just begun.

## 8.6 REFERENCES

Brown, R.A. (1976) A model of Jupiter's sulfur nebula. *Astrophys. J.*, **206**, L179–L183.

Carlson, R.W., Weissman, P.R., Smythe, W.D., Mahoney, J.C. and the NIMS science and engineering team (1992). Near-Infrared Mapping Spectrometer experiment on Galileo. *Space Sci. Rev.*, **60**, 457–502.

Frank, L.A. and Paterson, W.R. (2001) Passage through Io's ionospheric plasmas by the Galileo spacecraft. *J. Geophys. Res.*, **106**, 26209–26224.

Geissler, P.E., McEwen, A.S., Keszthelyi, L., Lopes–Gautier, R., Granahan, J., and Simonelli, D.P. (1999) Global color variations on Io. *Icarus*, **140**(2), 265–281.

Johnson, T.V., Veeder, G.J., Matson, D.L., Brown, R.H., Nelson, R.M., and Morrison, D. (1988) Io: Evidence for silicate volcanism in 1986. *Science*, **242**, 1280–1283.

Kieffer, S.W., Lopes–Gautier, R., McEwen, A.S., Keszthelyi, L., and Carlson, R. (2000) Prometheus, the wanderer. *Science*, **288**, 1204–1208.

Lopes, R.M.C., Kamp, L.W., Smythe, W.D., Mouginis-Mark, P., Kargel, J., Radebaugh, J., Turtle, E.P., Perry, J., Williams, D.A., Carlson, R.W., and Douté, S. (2003) Lava lakes on Io: Observations of Io's volcanic activity from Galileo NIMS during the 2001 fly-bys. *Icarus*, **169**(1), 140–174.

Lopes–Gautier, R., McEwen, A.S., Smythe, W., Geissler, P., Kamp, L., Davies, A.G., Spencer, J.R., Carlson, R., Leader, F.E., Mehlman, R., *et al.* (1999) Hot spots on Io: Global distribution and variations in activity. *Icarus*, **140**(2), 243–264.

Lopes–Gautier, R., Douté, S., Smythe, W.D., Kamp, L.W., Carlson, R.W., Davies, A.G., Leader, F.E., McEwen, A.S., Geissler, P.E., Kieffer, S.W., *et al.* (2000) A close-up look at Io in the infrared: Results from Galileo's Near-Infrared Mapping Spectrometer. *Science*, **288**, 1201–1204.

McEwen, A.S., Keszthelyi, L., Spencer, J.R., Schubert, G., Matson, D.L., Lopes–Gautier, R., Klaasen, K.P., Johnson, T.V., Head, J.W., Geissler, P., *et al.* (1998) Very high tempera-ture volcanism on Jupiter's moon Io. *Science*, **281**, 87–90.

Morabito, L.A., Synnot, S.P., Kupfermann, P.N., and Collins, S.A. (1979) Discovery of currently active extra-terrestrial volcanism. *Science*, **204**, 972.

Nelson, R.M. and Hapke, B.W. (1978) Spectral reflectivities on the Galilean satellites and Titan, 0.32 to 0.86 micrometers. *Icarus*, **36**, 304–329.

Peale, S.J., Cassen, P., and Reynolds, R.T. (1979) Melting of Io by tidal dissipation. *Science*, **203**, 892–894.

Rathbun, J.A., Spencer, J.R., Davies, A.G., Howell, R.R., and Wilson, L. (2002). Loki, Io: A periodic volcano. *Geophys. Res. Let.*, **29**(10), 10.1029/2002GL014747.

Sagan, C. (1979). Sulfur flows on Io. *Nature*, **280**, 750–753.

Smythe, W.D., Nelson, R.M., and Nash, D.B. (1979) Spectral evidence for $SO_2$ frost or adsorbate on Io's surface. *Nature*, **280**, 766.

Spencer, J.R., Stansberry, J.A., Dumas, C., Vakil, D., Prangler, R., Hicks, M., and Hege, K. (1997) A history of high temperature Io volcanism: February 1995 to May 1997. *Geophys. Res. Let.,* **24**(20), 2451–2454.

Wamsteker, W., Kroes, R.L., and Fountain, J.A. (1974). On the surface composition of Io. *Icarus*, **23**, 417–424.

Williams, D.A., Wilson, A.H., and Greeley, R. (2000) A komatiites analog to potential ultramafic materials on Io. *J. Geophys. Res.*, **105**, 1671–1684.

Witteborn, F.C., Bregman, J.C., and Pollack, J.B. (1979) Io, an intense brightening near 5 micrometers. *Science*, **203**, 643–646.

## Additional bibliography

*Galileo: Io up close*. Collection of papers in *Science*, **288**(5469), 1125–1288, 2000.

Cattermole, P. (1996) *Planetary Volcanism. A Study of Volcanic Activity in the Solar System* (Second Edition). Wiley–Praxis, Chichester, U.K.

Geissler, P.E. (2003) Volcanic activity on Io during the Galileo era. *Annual Rev. Earth Planet. Sci.*, **31**, 175–211.

Hanlon, M. (2001) The Worlds of Galileo: The Inside Story of NASA's Mission to Jupiter. St. Martin's Press, New York.

Harland, D. (2000) *Jupiter Odyssey*. Springer–Praxis, Chichester, U.K.

Lopes–Gautier, R. (1999) Volcanism on Io. In: H. Sigurdsson, B. Houghton, S.R. McNutt, H. Rymer, and J. Stix (eds), *Encyclopedia of Volcanoes*, Academic Press, San Diego, CA, pp. 709–726.

McEwen, A.S. (ed.) (2001) and papers therein: Geology and Geophysics of Io, special session in Journal of Geophysical Research vol 106, no.E12, pages 32,959-33,272.

McEwen, A.S., Kezthelyi, L., Lopes, R., Schenk, P., and Spencer, J. (2002) The lithosphere and surface of Io. In: F. Bagenal, W. McKinnon, and T. Dowling (eds), *Jupiter: Planet, Satellites and Magnetosphere*. Cambridge University Press, Cambridge, U.K., in press.

McEwen, A.S., Lopes–Gautier, R., Keszthelyi, L., and Kieffer, S.W. (2000) Extreme volcanism on Jupiter's moon Io. In: J. Zimbelman and T. Gregg (eds), *Environmental Effects on Volcanic Eruptions: From Deep Oceans to Deep Space*. Plenum, New York, pp. 179–204.

Morrison, D. (ed.) (1982) *Satellites of Jupiter*. University of Arizona Press, Tucson, AZ.

Rothery, D. (1992) *Satellites of the Outer Planets*. Claredon Press, Oxford, U.K.

Spencer, J.R. and Schneider, N.M. (1996) Io on the eve of the Galileo mission. *Ann. Rev. Earth Planet. Sci.*, **24**, 125–190.

# 9

# Ice volcanism on Jupiter's moons and beyond

*Louise Prockter* (Applied Physics Laboratory)

*I am a planetary geologist in the Space Department at the Johns Hopkins University Applied Physics Laboratory in Maryland. My fascination with icy satellites really took hold in graduate school. After completing my undergraduate degree in Geophysics at Lancaster University in England, I came to Brown University in Rhode Island to begin a PhD in planetary science, as opportunities to study the subject in my home country are rather limited. I arrived in July of 1994 and during my first week attended celebrations of both the 25th anniversary of Neil Armstrong's historic Moon walk, and the unbelievably serendipitous destruction of comet Shoemaker–Levy 9 as its pieces hurled themselves into Jupiter's atmosphere. This proved to be an auspicious (and fun) beginning to a career that continues to engage and fascinate me, and which I can never take for granted. I was fortunate enough to work on NASA's Galileo mission while in graduate school, studying Jupiter's icy moons Ganymede and Europa, and cutting my teeth on mission planning. After graduating four years ago I moved to the Applied Physics Laboratory (APL), where I spend about half of my time engaged in research on the surface geology of icy satellites and asteroids, and the remainder helping to plan and implement space missions. During my first year at APL, I worked on the Near-Earth Asteroid Rendezvous (NEAR) mission, which involved placing a spacecraft in orbit around the asteroid Eros, and ultimately setting it down safely on the surface. I am currently part of the team working on the imaging camera for the MESSENGER spacecraft, due to launch in 2004 on a journey to Mercury. I love my work, in which no two days are ever the same. I hope at some point in my career to help send a spacecraft back to Europa, surely one of the most fascinating and confounding places we have ever visited, and one with many secrets yet to tell.*

## 9.1   INTRODUCTION

Studying planetary images is a complex and sometimes frustrating process, in which the surface history of a body is commonly revealed only by teasing out information in tiny steps. Planetary geologists identify different surface features, or landforms, and look for relationships among them, trying to decipher which geological processes occurred, and when. This knowledge may then be extrapolated to areas of the planetary body that have not yet been investigated. Some planets, like Mars, are similar to our home planet and have, for the most part, recognizable features. Others, such as Jupiter's icy moon Europa, have bizarre landforms unlike anything seen on the rocky planets. The knowledge gleaned from the study of other planetary bodies brings an understanding to how our Earth came to be in its present state, and the possible types of worlds that could exist around other stars.

Ice-rich or "cryovolcanic" eruptions are any that consist of liquid or vapor phases of materials that freeze at temperatures of icy satellite surfaces, while cryo-magmatism is a more general term that includes rising diapirs of ice within the planet that are more buoyant than their surroundings. Despite being many hundreds of degrees cooler than their silicate terrestrial counterparts, cryolavas also have the power to sculpt landscapes and resurface worlds. Cryovolcanic substances may be water based, or might be comprised of other materials such as methane and ammonia. Depending on their compositions, cryolava viscosities can be frothy, runny, dense, and/or sluggish.

Several icy satellites show signs of resurfacing by icy flows, as evidenced by smooth plains and filled in or missing impact craters. Some exhibit features thought to form by the interaction of rising ice masses with the surface. Currently, active cryovolcanism is only known to occur on one solar system world, Neptune's moon Triton; but Jupiter's satellite Europa shows significant evidence of recent cryovolcanic activity, and may still be active today.

A significant hindrance to resurfacing with liquid water is that a column of water cannot rise above a less dense ice crust (hence ice cubes float). This is not an issue for molten silicate lavas, which are generally buoyant with respect to their cold, solid rock surroundings and can more easily migrate to the surface. There are several ways in which this density limitation can be overcome. If the surrounding crust is sufficiently contaminated by dense particles such as pieces of silicate rock, rising "clean" ice may be less dense and relatively buoyant. The buoyant ice may rise to the surface as diapirs, in a similar manner to salt diapirs on the Earth. Alternatively, water may be reduced in density by gas bubbles, such as those of contaminants like $CO_2$ or $SO_2$, allowing it to rise above surrounding water ice. Liquids can rise to the surface if they are overpressurized (squeezed) in isolated pockets or cracks beneath the surface. Another way to get fresh ice to the surface is if it is "warm" (i.e., ice that is close to its melting temperature and is thermally buoyant compared to its surroundings). Such ice may be relatively ductile, forming warm ice diapirs that may migrate upwards through the crust.

The presence and style of cryovolcanic processes occurring on a satellite depends on a number of factors that may include its size, location, composition, and orbital history. Larger satellites are more likely to have experienced heating during their initial accretion and differentiation, and may have undergone more radiogenic heating from the decay of radioisotopes in their complement of rock. The position of a satellite with regard to its parent planet and its sibling moons can determine whether it undergoes heating from being tidally squeezed. Changes in orbital eccentricity can result from periods of resonance, which raise tides on a moon and may pump sufficient energy to cause melting in its interior, leading to volcanism. This process is predicted on Europa (which is caught in a tug of war between Ganymede and Io), and may have occurred within other icy moons at different stages of their evolution.

Cryolava composition is predicted to vary depending on several factors, one of which is its location in the solar system. Theoretical modeling predicts that $H_2O$ ice should condense from the primordial solar nebula beyond Mars' orbit (the "frost" line), and that ices make up over half the mass of material which condensed from the nebula beyond $\sim 4\,AU$. Water ice has indeed been identified spectroscopically as the dominant constituent of the Galilean satellites, as well as those of Saturn and Uranus. Other possible constituents of cryolavas include methane ($CH_4$) and ammonia ($NH_3$), predicted to have condensed from the solar nebula at Jupiter and Saturn's distance from the Sun. In the far reaches of the solar system, carbon monoxide (CO), carbon dioxide ($CO_2$), and nitrogen ($N_2$) are expected.

The composition of a cryolava has a significant effect on its viscosity. At their melting points, solutions of brines are several orders of magnitude less viscous than fluid silicate lavas, so could potentially flood their surroundings, filling in topographic lows. In contrast, ammonia–water lavas are much "stickier", with viscosities closer to those of silicate lavas, and could construct volcanic topography similar to that observed on the Earth. Although predicted to be present in the Jovian system and beyond, ammonia is yet to be detected spectroscopically except perhaps on Pluto's moon, Charon.

A potentially important component of icy satellite volcanism may be clathrate hydrate, a crystalline phase of water ice in which other volatiles such as $CH_4$, $N_2$, CO, and noble gases are incorporated into solid cage-like structures in the water-ice lattice. Clathrate hydrates are well known on Earth, where they are found in such diverse environments as glacial ice, permafrost, and sea floor sediments. They have also been proposed to occur on Mars, where $CO_2$ clathrate hydrate may be a major constituent of the southern polar cap. Clathrates may be widespread in the outer solar system, and have been proposed to exist in cometary nuclei, Titan, Triton, and the Saturnian and Galilean satellites. When the temperatures or pressures of clathrates are raised above a critical value, the presence of the trapped volatiles may lead to gas-driven volcanism. For example, when methane clathrate is heated above $170\,K$, it breaks down to form ice and fluid or vapor. Such a process could cause explosive volcanism on icy satellites.

## 9.2   THE GALILEAN SATELLITES

In 1989, NASA launched the Galileo spacecraft from the space shuttle Atlantis, sending it on a 6-year journey toward Jupiter (e.g., Harland, 2000). Despite the failure of the spacecraft's main antenna to open, Galileo nevertheless sent back a wealth of data about Jupiter and its moons, studying their surfaces, atmospheres, magnetospheres, and charged particle environments (see Chapter 8). Galileo differed from its predecessors, two Pioneer and two Voyager probes, in that it was placed into orbit around Jupiter rather than just flying close by. The spacecraft's orbit allowed a series of encounters with Jupiter's four largest moons, as it flew past one of them roughly every few months, beginning in the Summer of 1996. Data of much higher quality and extent was obtained than was previously available, and images were returned by Galileo's camera, the Solid State Imaging System (SSI), at a range of resolutions from global (several km/pixel) to as high as 6 m/pixel. The images of the icy moons Europa and Ganymede along with data from the Galileo Near-Infrared Spectrometer (NIMS) enabled old (generally Voyager-based) models of cryovolcanic processes to be tested (Smith *et al.*, 1979; Carr *et al.*, 1995), and new ones to be proposed.

### 9.2.1   Ganymede

Ganymede is a world of opposites (Figure 9.1) (Pappalardo *et al.*, 2004). About one-third of the moon is very dark, and heavily pockmarked by impact craters of all

**Figure 9.1.** Left: Global view of Europa (radius 1,560 ± 10 km). Europa's bright icy surface is extremely young, and is marked with numerous linear ridges and cryomagmatic features. Center: Near-global view of Ganymede (radius 2,634 ± 10 km), the largest moon in the solar system. Large dark elliptical area in upper right is Galileo Regio, the largest piece of intact dark terrain. Bright areas are swaths of grooved terrain. Right: Global view of Callisto, which is almost as large as Ganymede (radius 2,400 ± 10 km) and has a heavily cratered surface similar to Ganymede's dark terrain.
NASA/JPL.

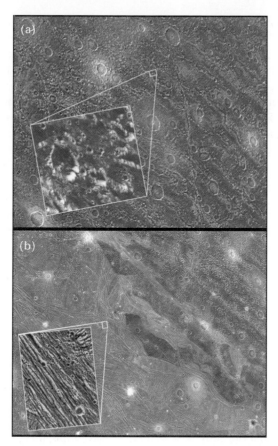

**Figure 9.2.** Grooved and dark terrain on Ganymede. (a) Typical dark terrain in Galileo Regio, which is heavily cratered and blanketed with dark material. Linear features are furrows, probably scars related to early giant impacts. (b) Typical groove terrain swath Uruk Sulcus; insert shows the closely packed ridges and troughs characteristic of much of the grooved terrain. (Scale: smallest discernable features are $\sim 80$ m.)
NASA/JPL/Brown University.

sizes, suggesting that it is nearly as ancient as the solar system itself. This landscape is termed "dark terrain" (Figure 9.2(a)). The other two-thirds of the moon are criss-crossed with vast swaths and polygons of younger bright material, some smooth, but most containing subparallel ridges and valleys (Figure 9.2(b)). The presence of this "grooved" or "bright terrain" indicates that Ganymede has at some time been resurfaced in a major fashion. The composition of the swaths—almost pure water ice—suggests that cryovolcanism was the likely culprit. What is not clear is when and how the resurfacing that created the grooved terrain took place, and why only two-thirds of the moon were affected, leaving large tracts of ancient dark terrain essentially untouched. Such knowledge would greatly increase our understanding of Ganymede's history and evolution, and with it, that of its sisters Callisto and

Europa. Ganymede appears to have been frozen in time midway between dark, heavily cratered Callisto on one side, and extremely young, bright, icy Europa on the other (Figure 9.1). Ganymede is postulated to have a liquid ocean, tens of km thick, although the ocean is currently thought to be about 170 km beneath the surface—too far to have any direct effect on surface features. Ganymede's ocean may be a remnant from a time when it was captured into one or more orbital resonances with Europa and Io, leading to tidal heating, internal melting, and enhanced geological activity.

While water ice is the predominant component of bright terrain, dark terrain appears primarily composed of hydrated minerals. Sulfur compounds are present, possibly implanted from the nearby moon Io, and some $CO_2$ is present, more abundant in the dark terrain than the bright terrain. Other simple organic molecules are inferred to result from radiolysis on the surface, or infall of meteoritic and cometary material. No evidence has been yet found for materials—such as $NH_3$—that could greatly reduce the melting point of cryolavas.

### Dark terrain

Dark terrain was inferred from low-resolution Voyager images to perhaps be blanketed by low-albedo lava flows, a model supported by crater counts that showed a dearth of small craters on some parts of the surface (a lava flow will fill in the smallest craters first). There were also suggestions that the rims of furrows— large ring systems probably formed around the sites of giant impacts (Figure 9.2(a))—were brighter than their surroundings and could be volcanic flows. A similar origin was proposed for light, apparently wispy patches of material on the surface. It was expected that evidence of cryovolcanism in dark terrain should be readily apparent in higher resolution images if it existed, either in the form of partially filled impact craters, or small volcanic cones. However, despite careful imaging and mapping of several areas of dark terrain, no unambiguous evidence of icy volcanism has been identified from high-resolution Galileo images. Either volcanism has not occurred, or has yet to be discovered. One reason why volcanic evidence may be hard to discriminate is that dark terrain is very old. An estimate of a planet's surface age can be obtained from the number of impact craters present, a technique that has been calibrated for the terrestrial planets using age-dating of Moon rocks, and for outer planets using the flux of potential impactors (e.g., comets; Zahnle et al., 2003). The large number of craters present on Ganymede's dark terrain suggests that it is extremely ancient—perhaps nearly 4.5 billion years—as it has been hit by numerous impacts during its lifetime. Features that are diagnostic of volcanism on other planets, such as source vents and flow fronts, may have long since been destroyed, or changed beyond recognition. Another possibility is that Ganymede's dark terrain has not yet been imaged at sufficiently high resolution to distinguish volcanic features, especially if they formed from low-viscosity lavas. The highest resolution images we have are sufficient to identify features no smaller than the size of an average house, so it is quite possible that we may not be able to see characteristic volcanic features such as flow fronts. Low-

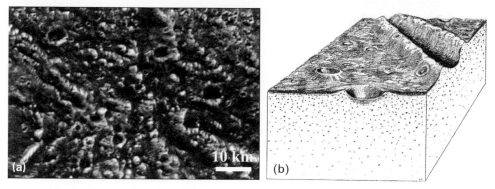

**Figure 9.3.** (a) High-resolution image of dark terrain in Galileo Region, showing dark streaks on slopes and dark material in troughs. (b) Sketch showing how dark terrain may be comprised of loose silicate lag material that has become concentrated on the surface of dirty ice below.
(a) NASA/JPL. (b) After Prockter *et al.*, 1998.

viscosity cryolavas could fill in any pre-existing topography, and may not undergo inflation events or form thicker stacks of overlapping flows, as are observed elsewhere, such as in Hawaii (see Chapter 2).

The Galileo images provide no definitive evidence that the dark terrain is formed from low-albedo volcanic flows. It is more likely that the terrain is composed of dirty ice with a dark (probably silicate-rich) lag deposit on its surface (Figure 9.3). This lag model idea was adapted for Ganymede from a "dirty snowball" model that was first proposed for comets in the 1950s. If a dirty snowball is left in the sun on an airless body, the ice will sublimate away until all that remains is a small pile of dirt. It appears that this process was, and perhaps still is, occurring on Ganymede; the dark terrain may be dirty ice, in which the surface has been heated by the sun allowing the volatile components to escape, and leaving a silicate "dirt" layer behind. The observations supporting this theory include what appears to be loose dark material in depressions such as valleys and crater floors, V-shaped streaks on slopes (a pattern consistent with mass wasting elsewhere in the solar system), and the slopes themselves, which are generally bright instead of dark (suggesting that dark material has been sloughing off) (Figure 9.3). Mass wasting would be aided by occasional shaking from nearby impacts and Ganymede-quakes. Although solar illumination is very weak at Ganymede, it is still sufficient to burn off volatiles in the relatively warm, dark terrain, especially given several billion years to do so. Therefore, it seems likely that volcanism is not necessary to explain the geological features found in the dark terrain, and no unambiguous evidence has been found for it to date.

### Grooved terrain

Ganymede's grooved terrain is much younger than its dark terrain, as evidenced by the lower density of impact craters present within it (e.g., Pappalardo *et al.*, 2004). Grooved terrain consists of swaths and lenses of icy material with surfaces spanning a range of textures from almost smooth, through subparallel ridges and troughs, to heavily cut up by orthogonal fractures (Figure 9.2(b)) (Pappalardo *et al.*, 1998). Much of the terrain is subdivided into different sized cells, and some of the ridge and trough terrain can be continuous for hundreds of kilometers. The ridges and troughs appear to have formed by extensional boudinage (or "necking"), or through domino-style normal faulting.

Early models suggested that the bright grooved terrain swaths formed when blocks of dark terrain were dropped down along faults, forming giant troughs, or graben (Figure 9.4(a)). These were thought to have subsequently become flooded with icy cryolavas of liquid or slushy water, forming smooth lanes. Some of the frozen swaths could then have undergone further tectonic expansion and/or shear deformation resulting in the formation of ridges and valleys. Galileo images of some smooth and highly linear swaths of terrain—informally termed "planks" (Figure 9.4)—have led to this model being revived. Detailed topographic models have been created using a combination of older Voyager data and new, higher resolution Galileo data (Schenk *et al.*, 2001). These 3-D models show that some of the smoothest swaths of bright terrain are also the youngest, and that these are topographically lower than the surrounding terrain. Older swaths tend to be more heavily grooved and tectonized, and sit topographically higher. These observations, along with images showing fine-scale embayment (onlapping onto or into surrounding terrain by a fluid) and burial of older features, suggest that the smooth planks formed when low-viscosity lavas flooded shallow troughs. After the lavas froze, extension and groove formation occurred.

One might wonder whether liquid water lavas such as those proposed for grooved terrain formation could flow any substantive distance on Ganymede's chilly, airless surface. The temperature ranges from a mere 90 to 160 K, as the moon receives only about 1/30th of the sunlight we enjoy here on Earth. Despite this extreme environment, lavas that are able to erupt could potentially flow for some considerable distance. The lack of atmosphere and correspondingly low vapor pressure means that any liquid erupting onto the surface will boil violently as soon as it is exposed to the vacuum of space. Such venting would explosively disrupt a rising magma into droplets, which would rapidly freeze and fall back to the surface as "cryoclastic" deposits. For small eruptions, a flow would probably freeze at its source. For eruptions with higher flow rates, the rapid boiling would form ice crystals at the surface that could be distributed throughout the flow. It has been suggested that a cover or "lid" of ice would rapidly form on top of the flow so that all boiling would cease. This would allow the liquid beneath to flow in the manner of silicate lavas, with a hot core and a chilled, insulating carapace. With sufficiently large discharge rates, and the presence of the cold brittle lid, a flow on Ganymede could potentially flood thousands of square kilometers before freezing.

A completely different kind of resurfacing model of the grooved terrain was proposed based on high-resolution Galileo images of severely tectonized areas, and the presence of close-spaced fractures in some areas of adjacent dark terrain. At high resolutions, the grooves are not quite so clean and bright as they may first appear, and have what appears to be loose, dark material in their troughs. Rather

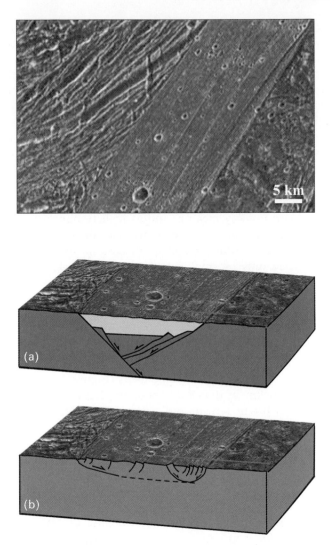

**Figure 9.4.** Arbela Sulcus, a relatively smooth linear "plank" within the Nicholson Regio dark terrain area. Two models for how such a smooth grooved terrain swath might have formed: (a) blocks drop down along faults, and the resulting trough becomes flooded with icy lavas, which may then themselves become faulted; (b) dark terrain is tectonized beyond recognition by numerous faults, leading to a relatively smooth appearance at current image resolutions.
Top image, NASA/JPL.

than originating as volcanic flows, the grooved terrain may instead be dark terrain that has been so heavily sliced up by tectonic processes that it is no longer recognizable (Figure 9.4(b)) (e.g., Pappalardo *et al.*, 2004). This "tectonic resurfacing" model does not require volcanism to have occurred, but neither does it preclude it. It is entirely possible that tectonic dismemberment of the terrain occurred after the bright swaths were cryovolcanically emplaced.

Despite the high-resolution images returned by Galileo, it is still not clear what are the relative roles of tectonic and volcanic resurfacing in creating grooved terrain, and the debate about how this bizarre landscape formed continues. There is ample evidence for shear along many grooved terrain swaths, and shear deformation is likely to be an important process in their formation. However, the exact interplay between tectonism and volcanism is not yet understood and continues to be studied.

As in the dark terrain, no evidence has been found in bright grooved terrain for flow fronts, constructional cones, or embayment relationships, all of which are diagnostic of volcanism elsewhere in the solar system. However, there are two features that are possible diagnostics of volcanism. Very smooth regions within the grooved terrain are notably lacking, and at high resolution almost all terrains appear to have undergone some degree of tectonic deformation. The smoothest imaged material is found within Harpagia Sulcus, in small areas up to 10 km across, and is disturbed only by small craters (Figure 9.5). It is possible that these areas formed by the distribution of fine material through seismic shaking from impacts, but it seems unlikely given the lack of potential source regions for such debris, or nearby slopes. Hence these smooth plains regions may be examples of icy volcanic flows.

The most probable cryovolcanic features identified anywhere on Ganymede are 18 or more arcuate depressions, or "paterae", found within the bright terrain, which could represent source vents for icy volcanic flows (Figure 9.6) (e.g., Pappalardo *et al.*, 2004). Averaging 50 km in diameter, these subcircular, scalloped features are always associated with bright terrain and some have central deposits that appear to have embayed their surroundings. One elongated patera has a deposit with arcuate flow-like morphology (Figure 9.6(b)), similar to ropey pahoehoe lava (see Chapter 2) although at a much different scale. This feature has a rim that stands up to 800 m above the surrounding terrain. The paterae may have formed in association with bright terrain, by the collapse of blocks over partially drained magma chambers in a similar manner to terrestrial calderas. This view is supported by Galileo images showing smooth interior deposits and possible evidence for collapse, giving rise to suggestions that the caldera-like features may be source regions for some of the bright terrain material. If these are the "smoking gun" of grooved terrain cryovol-canism, however, it is not clear why they are only found in some regions and are not widespread in distribution, as is the grooved terrain.

### 9.2.2   Europa

Europa's bizarre surface is crisscrossed with numerous double-crested ridges, wide smooth or ridged dark bands that cursorily resemble Ganymede's grooved terrain, and areas where the terrain seems to be smashed into giant iceberg-like plates. The surface is extremely young compared to other planetary bodies—the most likely estimate is an average of only ~50 million years—a mere blink of an eye in geological terms (Greeley *et al.*, 2004). The youngest impact crater is thought to have

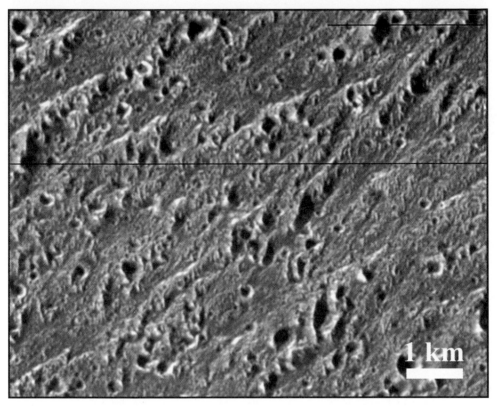

**Figure 9.5.** Harpagia Sulcus. This very high resolution view of a grooved terrain swath shows smooth material in-between small-scale ridges. The smooth material possibly has a cryovolcanic origin.
NASA/JPL.

formed within the last 20 million years, and parts of Europa may even be active today.

One of the most remarkable results from the Galileo mission is magnetic field evidence that implies the presence of a conductive layer—almost certainly liquid water—close to the surface (Kivelson *et al.*, 2000). Geological evidence suggests that the surface of Europa is probably decoupled from its interior, and may be "non-synchronously" rotating (i.e., the icy shell inching forward relative to the main body of the satellite). Although it is not known exactly how fast the shell is moving, there must be a liquid or ductile layer separating the shell from the interior for this process to occur. If there is a liquid water ocean beneath Europa's icy shell, then it may be a place conducive to the formation of life as we understand it from our terrestrial viewpoint. Life on Earth is found wherever water is found, and the search for water on Mars has been a priority for this very reason. Thus Europa is of great interest to astrobiologists, particularly those who study life in extreme environments.

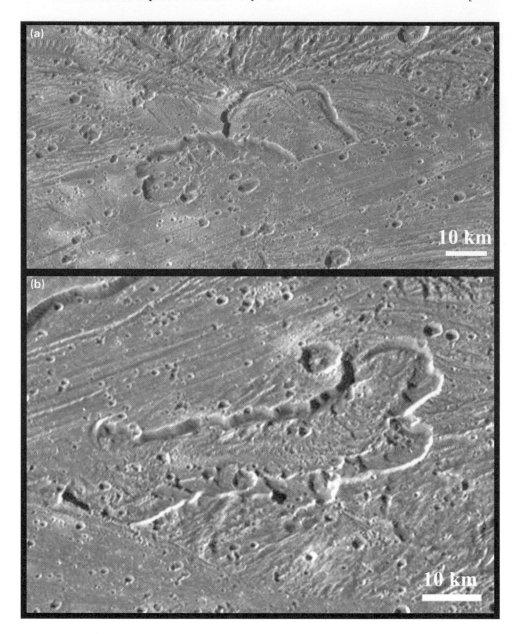

**Figure 9.6.** Caldera-like features within grooved terrain. (a) Two scalloped features with clearly defined margins appear cross cut to the lower right by a younger smooth swath. (b) Elongated scalloped feature with central deposit that may have flowed to the lower left. Surface of the center deposit has a ropey texture reminiscent of terrestrial pahoehoe lava flows. NASA/JPL.

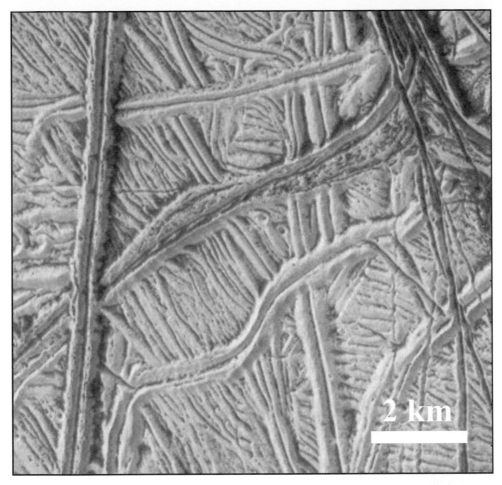

**Figure 9.7.** Examples of double ridges on Europa's surface. Ridges have two uniform crests alongside a central trough, and are commonly linear in planform.
NASA/JPL.

## Ridges

Europa's enigmatic double ridges are its most ubiquitous landform, but their origin is not well understood. The majority of these ridges consist of two uniformly matched crests separated by a V-shaped trough (Figure 9.7), and they can continue for hundreds of kilometers across the surface. Some ridges have a cycloid form. This has been suggested to result from progressive cracking of the ice due to daily tidal stressing of Europa's shell as it orbits Jupiter. Most ridges are remarkably straight, and may have formed by faster cracking.

Several models have been proposed for the formation of the ridges, in particular their raised crests, and some of these have a magmatic or cryomagmatic origin (see review in Greeley *et al.*, 2004). One model suggests that the ridges crack above thermally or compositionally buoyant linear ice diapirs, the rising of which causes upwarping of the crack margins to form ridge crests. This model does explain the morphological features observed along ridges, but it is difficult to explain how this process would account for the great length of some of the ridges. Another model advocates that the ridges form in a similar way to linear features found in terrestrial sea ice. Ice ridges called "leads" on Earth form when ice cracks and moves apart, then comes back together crushing any thin ice and ice crystals that have formed in the temporary gap. Continued pumping of water and ice into the gap in this way results in the build-up of ridge crests. This model requires that cracks remain open all the way to an ocean very close to the surface, for which there is little evidence on Europa. Another model suggests that the ridges form through compression, but although there is evidence that some ridges may have a compressional origin, it is difficult to explain the presence of the well-developed trough between the ridge crests. It has been suggested that Europa's ridges form from linear volcanism, in which the ridge crests are the product of gas-driven fissure eruptions. The main drawbacks with this model are the difficulty in keeping cracks open for sufficient time to form the uniform ridge crests, and the improbability of maintaining volcanic eruptions over the extreme length of the ridges.

Perhaps the most widely accepted model for ridge formation at the current time is one in which the ridge crests form due to shear heating along cracks. Tidal stressing causes the cracks to slip forward and back by small amounts, and the resulting friction between the two crack faces may cause melting and buoyant rising of "warm" ice. This model may account for the intrusion and uplift proposed in the linear diapirism model, and can explain the length of the ridges.

The only other place in the solar system in which Europa-style ridges have been found to date is Triton, and it is likely that the ridges on that moon, although of a larger scale, may form in a similar fashion.

### Bands

Many of Europa's landforms were created when new material was emplaced onto the surface from below, either in liquid form, or as buoyant, ductile ice. Bands can be hundreds of kilometers in length, and formed when the surface split and moved apart, allowing new material to come up from below (Figure 9.8) (e.g., Greeley et al., 2004). Unlike Ganymede's grooved terrain, we know that Europa's bands are formed of completely new material, because the edges of the bands can be reconstructed, allowing pre-existing lineaments to line up almost exactly. The edges of the bands probably separated along pre-existing faults, due to stresses caused by Europa's orbital motion around Jupiter and the movement of the icy shell relative to the satellite's tidal axes as a result of its non-synchronous rotation. The stresses produced by this motion result in the formation of numerous cracks in predictable orientations across the surface. Sometimes the stresses are sufficient to pull a crack apart, allowing new material to well up into the gap from below.

The detailed morphology of band interiors is strikingly similar in many respects to that observed along the Earth's mid-ocean ridges. On Earth, these primary sites of

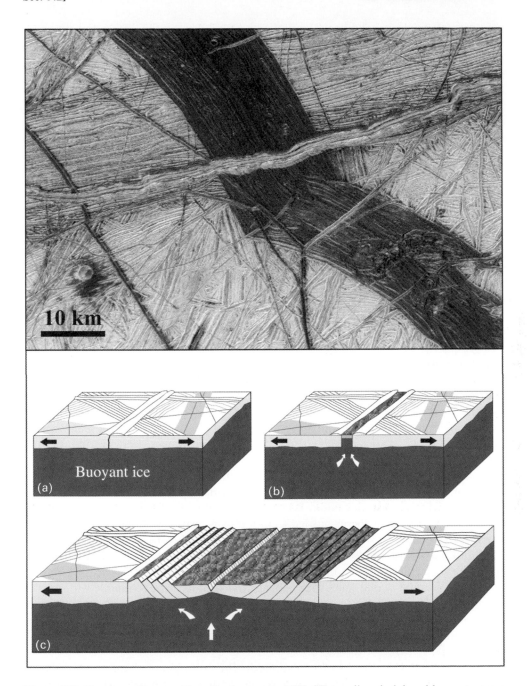

**Figure 9.8.** Bands on Europa. Top: Dark cuspate NW–SE-trending dark band has cut across an older gray band trending E–W. Bottom: diagram showing possible method by which bands form, in a similar manner to terrestrial sea floor spreading.
After Prockter *et al.*, 2002, copyright AGU/JGR.

new crust formation have broadly similar textures across the globe, but at small scales exhibit differences depending on characteristics including the speed at which the ridge spreads apart. Europa's bands have several features analogous to those found on terrestrial mid-ocean ridges, including fields of small cryovolcanic hummocks, subparallel faults, and well-developed medial troughs. They appear to form in a similar way, spreading symmetrically from a central axis. Unlike mid-ocean ridges, bands do not continue to spread indefinitely, but rather reach some critical width of about 25 km before ceasing to open.

By studying the morphology of the bands, it becomes clear that they did not form from liquid water, but must have been emplaced as solid (although ductile) ice. Several observations lead to this conclusion, including the presence of topographic features such as small hummocks. The band material must have been solid during the formation of the hummocks, since liquid water cannot support such structures. In addition, stereo modeling of some bands shows that they stand hundreds of meters above the surrounding terrain, implying a solid, rather than liquid origin. Finally, there is no evidence for embayment of surrounding ridges and valleys bounding the band, something that would be expected had they been liquid at the time of their formation.

### Chaos and lenticulae

Bands are not the only cryovolcanic features that have drastically altered Europa's surface. Large areas of terrain appear to have been smashed to pieces, and are called, appropriately, chaos regions. Chaos consists of plates and blocks of the pre-existing icy surface that appear to be embedded within a darker and finer-textured matrix material (Figure 9.9(a)). Some of the plates can be reconstructed like a giant jigsaw puzzle, but much of the original surface is either no longer recognizable, or has completely vanished. The chaos regions, and their smaller counterparts, lenticulae, have caused some of the most lively debates within the Europan community, because of the implications of different models of their formation for the thickness of the ice shell (see reviews in Greeley et al., 2004). The visual similarity between some chaos terrain and terrestrial icebergs has led to models suggesting that chaos formed when liquid water melted its way through a thin (< 6 km) ice shell to the surface, breaking it up and allowing some blocks to melt. Other models argue against such an origin on the basis that it would require an impossibly large amount of tidal heating energy to be concentrated in one small spot beneath the shell in order to enable widespread melting to take place. Further evidence against the "melt-through" model is that some of the chaos regions are known to stand several hundred meters above the surface, implying a solid ice origin since denser water would not be able to rise this high (Figure 9.9(b)).

Currently, the most widely accepted model for the origin of chaos terrains is that they formed through the interaction of rising ice diapirs with the surface of a relatively thick (15–30 km) shell. Such "warm" ice need only be close to its melting temperature to be sufficiently buoyant to rise up through a colder ice shell. Depending on the thermal properties and velocity of a rising ice diapir, the

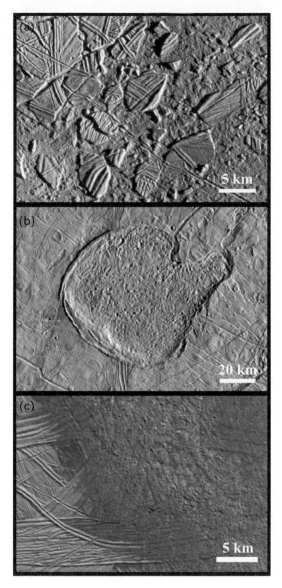

**Figure 9.9.** Cryovolcanic chaos features on Europa. (a) Conamara Chaos, an area in which plates of pre-existing terrain have moved and rotated in a matrix of lower albedo material. (b) Murias Chaos, a different morphological type of chaos in which the pre-existing terrain is barely recognizable, leading to a texture like frozen slush. This feature bulges several hundred meters above the surface and has downwarped the terrain on the western side, while apparently overflowing across the eastern boundary. (c) The western margin of Thrace Macula, where chaos appears to have formed by *in situ* disintegration, as evidenced by the faint traces of pre-existing ridges. Dark material has flowed away from the chaos region to embay troughs within the surrounding terrain, suggesting a low viscosity.
NASA/JPL.

overlying terrain could either break up into plates, or could just disaggregate *in situ* forming matrix material (Figure 9.9(c)). A variation on this model is that chaoses formed through a combination of rising warm ice interacting with pockets of low-melting-point brines that have been trapped in the ice close to the surface. This theory could account for the different heights of the blocks within the matrix, and the occasional presence of dark frozen fluid embaying surrounding troughs (Figure 9.9(c)).

The ice diapir theory is consistent with the size and morphology of lenticulae, small subcircular features that include pits, dark spots, and domes, many of which have surfaces that are broken up like chaos (Figure 9.10). These features are numerous but cluster around ∼10 km in diameter, precluding an impact or melting origin. The size distribution of lenticulae argues strongly in favor of a diapiric origin: similar-sized diapirs are predicted to form when an ice shell thickens to a critical degree at which solid-state convection can be initiated. For an average lenticula size of 10 km, this would imply an ice shell of at least comparable thickness. It has been suggested that lenticulae and chaos are essentially the same genetic feature as manifested at different scales, or that the chaoses form when several lenticulae erupt onto the surface close to one another and coalesce.

### Dark plains

Some apparently fluid cryolavas have extruded onto the surface of Europa, from unknown depths. Figure 9.11 shows examples of apparently frozen pools of material. The cryolavas probably had a relatively low viscosity when emplaced, as evidenced by the embayment of surrounding troughs. It would be more difficult to create this type of relationship if the erupted material had been very viscous. The dark material apparently oozed in between the ridges before freezing.

The frozen lavas have a reddish brown color, indicating that they contain some dark, non-ice contaminant. Data from the Galileo Near-Infrared Mapping Spectrometer (NIMS) suggest that the reddish brown areas contain hydrated salts. Although the salts are not themselves colored, they are clearly associated with recently emplaced material, which might include colored sulfur compounds. Such cryolavas may have been transported from a brine-rich sublayer, possibly the ocean.

Dark material filling topographic lows is also present in and around some chaos and lenticulae (e.g., Figure 9.10(a)). This material has been suggested to be the surface expression of brine pockets that have been melted by upwelling ice diapirs. Such pockets of salty "antifreeze" are proposed to have been located very close to the surface, and to have partially melted when warmed by rising ice diapirs.

### "Cryoclastic" eruptions

Pyroclastic eruptions are common on terrestrial planets, and may have occurred in a frozen, or cryoclastic form on Europa. Evidence for cryoclastic eruptions may come from "triple bands"—bright single ridges with dark material along their flanks (Figure 9.12). One unusual triple band, Rhadamanthys Linea, has dark spots spaced along its length, rather than contiguous dark deposits (Figure 9.12). The

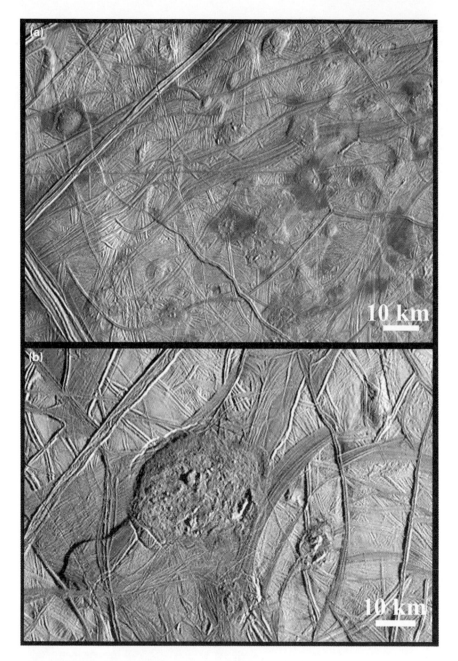

**Figure 9.10.** (a) A number of lenticulae with a variety of textures, including pits, spots, domes, and micro-chaos regions. Their close spacing and similar size distribution implies an origin as cryomagmatic diapiric features. (b) High-resolution view of a lenticula which has broken the surface into fine-scale chaos.
NASA/JPL.

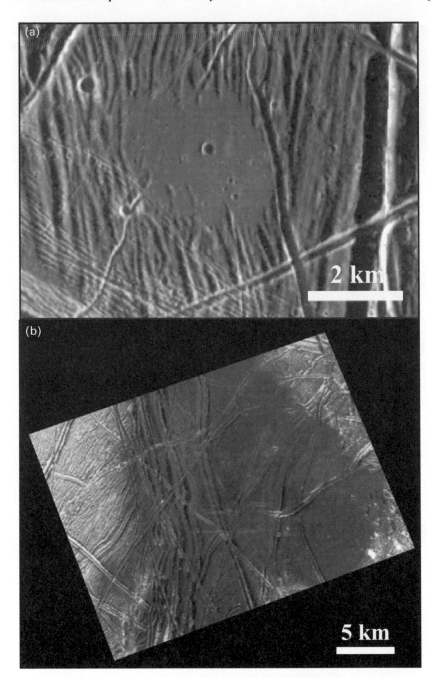

**Figure 9.11.** Dark plains material. (a) A fluid origin is suggested by this feature's smooth, apparently flat surface, and the embayment of surrounding troughs. (b) Castalia Macula, a dark plains deposit that may have buried several ridges.
NASA/JPL/PIRL, University of Arizona.

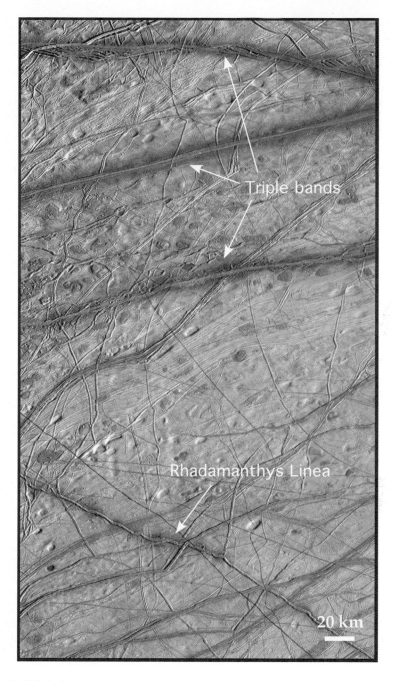

**Figure 9.12.** Triple bands, each consisting of a central bright ridge with dark margins on either side. Rhadamanthys Linea is atypical, with dark material in spots or patches along its length.
NASA/JPL

dark spots are thought to result from fissure eruptions, which begin along a narrow crack and may become concentrated in a few places as they mature (Fagents *et al.*, 2000). It is evident from Galileo images that this dark diffuse material drapes over pre-existing topography like a thin mantle, implying a particulate, rather than a fluid origin. The spots are likely the result of explosive venting and fallout of cryoclastic material. The source material may have been a body of water containing dissolved volatiles such as CO, $CO_2$, or $SO_2$, which would have exsolved as they approached the surface and became depressurized, accelerating as they did so (this is the same process that occurs if a bottle of soda is shaken then immediately uncapped) (see Chapter 10). Some mechanisms by which the necessary rapid decompression in cryomagmas may have occurred include the upward propagation of cracks from the interface of a water layer and an icy crust, rise from pressurized cryomagma reservoirs or intrusions, or rising ice plumes. Once a stream of gas and liquid hits the vacuum at the surface, it is blasted out in ballistic trajectories due to Europa's small gravity and the lack of an impeding atmosphere. Some of the ballistically ejected matter would fall to the surface forming circular deposits around its source vent.

### Stratigraphy

In order to understand the history of a region, geological maps are constructed showing the relationships between different features. These are then converted into stratigraphic columns, showing the relative ages of landforms. Overlapping relationships among features on Europa's surface show that chaos and lenticulae sit very high in the stratigraphic column (i.e., they are very young compared to the rest of Europa's surface) (e.g., Greeley *et al.*, 2004). Bands are generally intermediate in age, and predate the chaos units. The apparent change in the style of surface processes over time has led to a theory that an ocean below Europa's surface is slowly freezing, so that the icy shell is thicker than it once was. The formation of bands is consistent with lateral tectonics within a thinner ice shell, in which a ductile ice layer used to be closer to the surface and welled up when a gap appeared in the overlying ice. If the ocean slowly froze out since that time, resulting in a thicker ice shell, it would be more difficult for bands to form. Eventually the ice could have reached a sufficient thickness for the initiation of convection, which requres a thicker shell, and which could explain the chaos and lenticulae are now present on the surface. Other evidence for a thicker (>15 km) ice shell on Europa comes from impact studies and topographic modeling. For example, some of Europa's impact craters have distinct central peaks that would only have formed if the impact had occurred in a thick ice shell—thin ice would not be able to support such topography and a flatter crater would be expected.

### 9.2.3  Callisto

Callisto's ancient surface is very similar in appearance to Ganymede's dark terrain, and is characterized by impact craters (Figure 9.1(c)) (Moore *et al.*, 2004). Early low-resolution images returned by the Voyager spacecraft hinted at dark, smooth areas within the cratered plains, and these were thought to be good candidates for cryo-volcanic deposits. Higher resolution Galileo imaging has found no evidence for this, however, and it seems that the plains are the result of mass wasting and sublimation of ice from the surface. As with Galileo images on other Jovian moons, it is possible that we do not have sufficient coverage at high enough resolution to resolve cryo-volcanic features on Callisto, or that the surface is so old that they have become degraded beyond recognition.

## 9.3   THE SATURNIAN SATELLITES

After leaving the Jovian moons, Voyager 2 sped past the Saturnian system in 1981, obtaining images of several of Saturn's moons (Figure 9.13(a)) (Smith *et al.*, 1982;

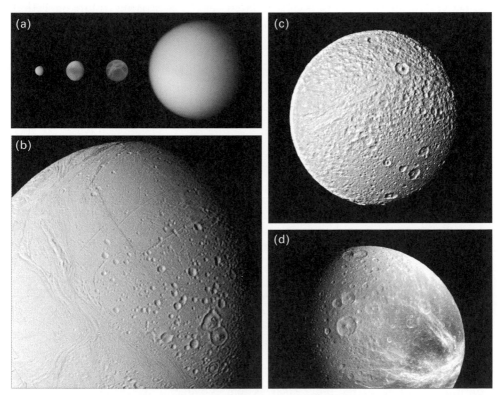

**Figure 9.13.** Saturnian satellites which show evidence of resurfacing, possibly by cryovolcanic processes. (a) From left: Enceladus, Tethys, Dione, and Titan, to relative size. (b) Enceladus, showing ridge belts and smooth plains, possible evidence of volcanic resurfacing. (c) Tethys, showing marked variations in crater density across its surface, an indicator of probable cryovolcanic resurfacing. (d) Dione, showing both heavily cratered and relatively smooth regions, suggestive of partial resurfacing by icy lavas.
Photo montage by P. Schenk, LPI, Images, NASA/JPL.

see review in McKinnon, 1999). At the time of writing, the Cassini spacecraft has just begun to explore Saturn and some of its moons, perhaps revolutionizing our ideas about them as the Galileo spacecraft has for the Jovian satellites.

Spectra of the Saturnian satellites show that, with the exception of Iapetus, they are comprised of remarkably clean water ice. Enceladus, although only ~ 500 km across, shows clear evidence of cryovolcanic resurfacing in recent geological times, and perhaps continuing at present. Part of the satellite is cratered, but large regions are smooth and relatively crater-free, containing faults and perhaps folds (Figure 9.13(b)). At least one of these faults has been dropped down, then possibly flooded with fresh ice. The surface of Enceladus is brighter than that of any other solar system satellite, which implies that parts of it are very fresh and may be covered with recent bright frost deposits. Tethys is much larger (1,060 km in diameter) than its brighter sibling (Figure 9.13(c)). A high density of impact craters on some parts of its surface show that they are ancient, while other regions with fewer, and generally smaller craters appear to have been resurfaced. The most striking feature on Tethys is the huge valley Ithaca Chasma, which is up to 100 km wide in places and stretches at least three-quarters of the distance around Tethys' circumference. Dione is similar to Tethys in size (1,120 km in diameter) and has similar surface features but exhibits more extensively resurfaced areas (Figure 9.13(d)). This resurfacing did not take place recently, as the areas have many superposed impact craters. Dione also has linear features and troughs, which probably have a tectonic origin.

The lavas responsible for the cryovolcanic flows inferred on the surfaces of these three satellites may have had a composition of ammonia and water. Ammonia may have been bound in the water-ice crystal lattice of the satellites as a hydrate, when temperatures in Saturn's protosatellite nebula cooled to about 150 K. Ammonia acts as a highly efficient antifreeze—it can suppress the freezing point of an aqueous solution by almost 100 K. Thus, when ammonia is present along with ice, only moderate increases in temperature are necessary for melting to occur in a satellite. At 176 K, an ammonia-rich melt can form a mixture that is slightly buoyant with respect to surrounding unmelted water ice, aiding rise and eruption. As yet, however, ammonia has not yet been identified on any solar system satellites, possibly because its molecules may not survive long in a vacuum and within the harsh radiation environments present at these moons. The Cassini spacecraft may be able to identify ammonia if it is present in young areas such as around fresh impacts.

## 9.4    THE URANIAN SATELLITES

Continuing its tour of the outer solar system, Voyager 2 passed close to Uranus' southern hemisphere in 1986, obtaining only a small number of satellite images as it passed by (Smith *et al.*, 1986; see review in McKinnon, 1999). The Uranus encounter was designed to slingshot the spacecraft toward its final rendezvous with Neptune, but as a result, Miranda was the only satellite for which detailed images were obtained. The 4 major satellites imaged more distantly by Voyager show some evidence of resurfacing (Figure 9.14). A range of compositions and eruption con-

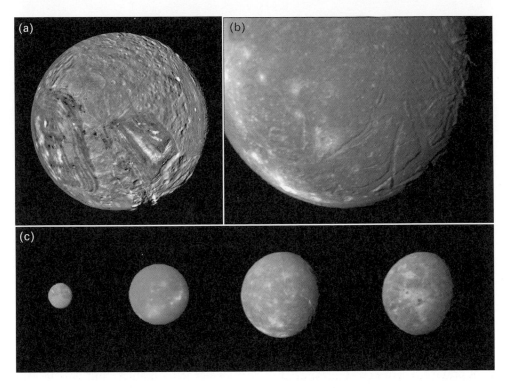

**Figure 9.14.** Saturnian satellites which show probable evidence of resurfacing. (a) Miranda, which has ovoidal areas of volcanic and tectonic resurfacing, possibly the result of internal heating and melting of low-temperature ices which have impinged upon the surface as diapirs. (b) Ariel has vast extensional canyons and associated smooth plains indicative of cryovolcanic lava flows. (c) In addition to Miranda (left) and Ariel (center left), Titania (center right) and Oberon (right) also show some signs of cryovolcanic activity. Titania has a lack of large craters in some areas, suggesting early resurfacing, while Oberon has dark material in some of its craters, perhaps dark icy extrusions related to the impacts.
Photo montage by P. Schenk, LPI., Images, NASA/JPL.

ditions are inferred from the variety of observed morphological types and albedo variations. The most common type of flows are flood deposits, which may have erupted from fissures. Small circular vents and evidence of some cryoclastic activity have been postulated on Ariel and Miranda.

Miranda's surface, like Ganymede's, appears frozen partway between two geological states (Figure 9.14(a)). Much of the satellite's surface is composed of bright cratered terrain, but embedded within this lie three vast, dark polygonal features termed coronae. They are interpreted to have formed through uplift, which led to extension and faulting, along with extrusion of cryomagmas through fissures. These icy lavas were probably viscous enough to form ridges within the coronae. This catastrophic partial resurfacing is likely the result of the beginnings of differentiation in this tiny (500 km diameter) moon, an event that was not able to progress to completion.

Ariel (1,160 km in diameter) exhibits a large number of fault-bounded canyons, many of which are associated with smooth flows, particularly within their floors (Figure 9.14(b)). The presence of sinuous troughs along the centers of these probable flows, and their arched cross sections, imply that the emplaced material was relatively viscous. Titania (1,580 km in diameter) also shows definitive evidence for extension across its surface, in the form of distinct fault-bounded valleys up to 1,500 km long and 100 km wide. A lack of large craters suggests some widespread resurfacing early in its history (Figure 9.14(c)). Titania's relatively rock-rich composition would have supplied a significant amount of radiogenic heating early in its history, but perhaps only enough for modest differentiation to occur. Oberon (1,525 km in diameter) was poorly imaged, but some dark patches are visible within the craters (Figure 9.14(c)). These may have formed when icy lavas containing a dark contaminant (perhaps rich in carbon) were extruded onto the surface, but otherwise no evidence of cryovolcanic resurfacing is apparent on this moon.

## 9.5  NEPTUNE'S SATELLITE TRITON

Triton (2,705 km in diameter) is Neptune's only large moon, and the only one of its moons to be imaged in sufficient detail to discriminate morphological features other than craters. The moon is comprised of rock, water ice, organics, and volatile ices (including $N_2$ and $CH_4$) (e.g., Cruikshank, 1999). Voyager 2 explored Triton in 1989, returning images of over half of the surface, $\sim 20\%$ of which were at 1 km/pixel or better resolution. These revealed a wealth of geological features on a surface that has few impact craters and appears remarkably young, and may, like Europa, still be active today. Many landforms are interpreted to be cryovolcanic in origin, including eruptive hills, flow lobes, caldera-like depressions, the enigmatic "cantaloupe" terrain, and what may be extensive pyroclastic sheets. Superposition relationships among different types of features hint at multiple episodes of both explosive and effusive events, and some flows have thicknesses that imply rheologies comparable to terrestrial andesite, dacite, or rhyolite. The explosive nature of some of the landforms implies that significant quantities of dissolved gases were present in their source magmas.

Modeling of Triton's resurfacing rate based on the paucity of its craters suggests that it may be second only to Io and Europa in its level of geological activity in the outer solar system (Stern and McKinnon, 2000). The reason for this apparent activity is that Triton was probably a Kuiper Belt object that was originally in an independent orbit around the Sun. At some time in its early history, it was captured by Neptune's gravity, a cataclysmic event that caused widespread melting and differentiation of the new satellite along with outgassing of a massive atmosphere. Tidal heating may have sustained warm interior temperatures for upwards of a billion years, and a subsurface liquid ocean may still persist today if ammonia is present in the icy mantle.

*Cryovolcanic features*

Triton has many landforms that resemble terrestrial volcanic features (Figure 9.15). Some of the most prominent are small circular to elongate cones, about 7–15 km in diameter. Their summits commonly contain smooth sided pits typically 4–7 km in diameter, which can breach the side of the cone. Cones may occur individually or in clusters, and chains of pitted cones are common. The cones may be analogous to cinder cones, and even though they are substantially larger than cinder cones on the Moon and Mars they have similar dimensional ratios suggesting that they may be formed of particulate material at the angle of repose. Chains of pits along ridges are similar to terrestrial tectonically controlled cinder cones and explosion pits, and may have a similar origin as cryovolcanic constructs along fractures, with slopes of smooth cryoclastic debris. Pit paterae are circular to elongate depressions typically 10–20 km in diameter which can have raised rims and are generally located within patches of smooth material 100–200 km in extent. The paterae occur singly or in chains. Ring paterae are much larger, 50–100 km in diameter, and have an outer rim

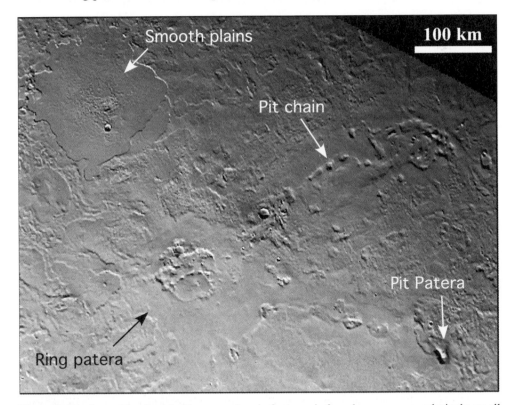

**Figure 9.15.** A wealth of probable cryovolcanic features is found even over a relatively small region of Triton. This view shows smooth plains both filling shallow circular depressions and forming a "ring" patera, chains of volcanic pits, and isolated pits, all of which are similar to volcanic features on terrestrial planets.
NASA/JPL.

defined by a ring of coalescing pits. The ring and pit paterae and associated smooth deposits are interpreted to have formed as explosive cryovolcanic craters and cryoclastic deposits. The smooth plains surrounding these features locally show contacts with the surrounding terrain suggestive of cryoclastic deposition (i.e., they mantle, rather than embay the surrounding topography, and their margins can "feather" out into the surrounding terrain). Chains of pit paterae follow tectonic trends in the region and may be analogous to chains of explosion and collapse pits in linear volcano-tectonic zones such as those in Iceland (see Chapter 10).

Some of the most bizarre features on Triton's surface are the guttae (Figure 9.16), huge dark lobate features 100–200 km across each surrounded by a bright aureole of fairly constant width (20–30 km wide). The smooth surfaces and lobate edges of the guttae suggest extrusive materials that have flowed in a viscous manner. Guttae deposits have an inferred thickness of at least tens of meters, so if they are extrusive, they must have had relatively high viscosities when they were emplaced. The origin of the guttae aureoles is not known: suggestions include low-viscosity flows, condensed gases, or thermally metamorphosed surface materials.

Triton has a significant number of vast ridges, measuring $\sim$15–20 km in width, and in some cases continuous for $\sim$800 km (Figure 9.17). Slightly sinuous, they are steep-sided in profile and some appear to stand above the level of the surrounding terrain. A deep continuous axial depression $\sim$5–10 km wide results in a double-ridge morphology, and some ridges are bounded by shallow troughs $\sim$20 km wide. The ridges appear to span a range of ages and degradation states and are proposed to have formed by the eruption of viscous material into the axes of graben, which then built up as the degree of extension and/or magma availability increased.

Triton's so-called "cantaloupe" terrain contains cavi, quasi-circular shallow depressions typically 25–35 km in diameter with slightly raised rims, giving the uncanny appearance of cantaloupe melon skin (Figure 9.17). Cavi interiors can be smooth, rough, pitted, or ridged, and some contain smooth, lobate deposits. The cavi have been suggested to represent cryovolcanic explosion craters, such as terrestrial maars, on the basis of their similar morphologies. An alternative mode of formation is by diapirism: the organized cellular pattern of the cantaloupe terrain has been proposed to closely resemble terrestrial salt diapirs, and the terrain may have formed due to density inversion in a layered crust composed partly of ice phases other than water ice. They may be analogous to Europa's lenticulae, which have also been proposed to have a diapiric origin.

### Plumes

Voyager 2 images of Triton showed suspicions of active surface processes in the form of dark streaks superimposed on the bright southern polar cap (Figure 9.16; see also Chapter 11). Although the presence of dark material was not in itself remarkable—for example, $CH_4$ on the surface can be converted to dark organic material by a variety of energetic processes—what was surprising was the fact that the streaks were visible at all. Triton is expected to undergo a cycle of volatile sublimation and deposition from pole to pole on the cycle of 1 Triton year ($\sim$165 Earth years),

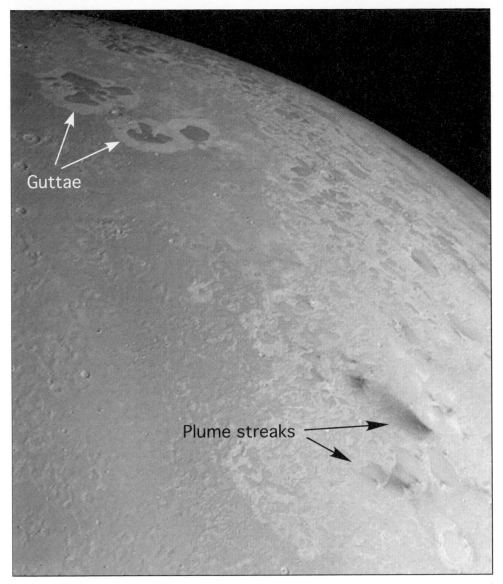

**Figure 9.16.** Triton's mysterious guttae are giant lobate features, each with a dark core and bright annulus. Dark plume streaks overlie the bright south polar cap. (Image is ~1,500 km across.)
NASA/JPL.

resulting in a meter or more of $N_2$, $CH_4$, and $CO_2$ frosts being removed from one pole and deposited on the other. The presence of the dark streaks on the surface implied that they were probably younger than a Triton year. It was not until shortly after the Triton encounter by Voyager, however, that the cause of this "smoking

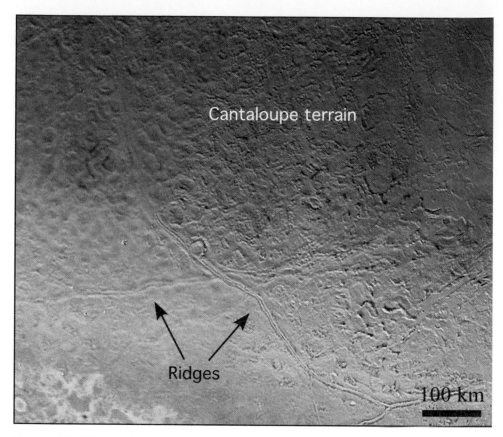

**Figure 9.17.** Triton's bizarre "cantaloupe" terrain has ridged and dimpled skin suggestive of an origin by cryomagmatic diapirs. Ridges found across the surface have similar morphologies to those on Europa.
NASA/JPL.

gun" was identified; stereoscopic images used to search for material above the illuminated disk showed clear plumes, probably composed of dust and gas (Kirk *et al.*, 1995). The plumes appeared as dark columns up to 8 km in height, with estimated radii of up to ~1 km, likely composed of fine dark particles and/or condensed volatiles and vapor. The columns fed long clouds of dark material that drifted with the tenuous winds for more than one hundred kilometers.

One model that has been suggested as the mechanism driving the plumes is that of explosive venting of nitrogen gas pressurized by solar heating. The surface of Triton is approximately 38 K, one of the lowest surface temperatures in the solar system, and appears to be blanketed with a relatively thick ($\geq 1$ m) layer of transparent solid $N_2$. This model proposes that dark material at the base of the layer is warmed by sunlight, and undergoes a significant increase in temperature with respect to the surface. The warming results in a substantial increase in the vapor pressure of the $N_2$. If this highly pressurized $N_2$ vapor is first trapped in pore spaces, then released to the surface through a vent, the resulting geyser could entrain subsurface dark material, lofting it into the atmosphere and accounting for the dark streaks associated with the plumes. A temperature increase of only 2 K would be sufficient to propel the plumes above the surface to the observed altitudes. An alternative driving mechanism suggests that the heat source for the geysers comes not from the Sun, but from within the satellite. If the solid-$N_2$ polar caps or subsurface ice are undergoing thermal convection in addition to conduction, heat is transported to the surface and could result in a temperature rise sufficient to drive the geysers. Although other models have also been proposed, an explosive venting model seems the most likely, regardless of the heat source.

## 9.6  PLUTO

To date, Pluto and its close moon, Charon, have been imaged in only limited detail by ground-based observations and the Hubble Space Telescope (Figure 9.18) (see review in Cruikshank, 1999). The smallest of the planets—it has a diameter of only $\sim 2,300$ km—Pluto's surface composition is 98% $N_2$, with some $CH_4$ and traces of CO. Images of the surface reveal large light and dark areas, generally with lighter regions near the poles, and darker surfaces near the equator. Because of its composition and its highly elliptical orbit, it has been suggested that Pluto represents a remnant planetesimal that has survived from the early solar system and may be best grouped with objects of the Kuiper Belt. The similarity of bulk properties of Pluto with those of Neptune's moon Triton has led to suggestions that Triton, like

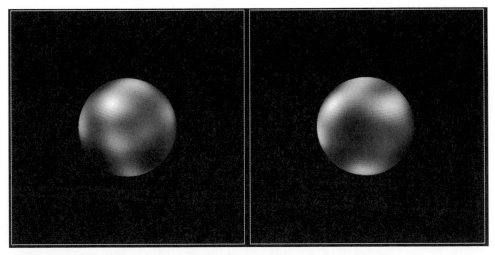

**Figure 9.18.** Hubble Space Telescope images of the two hemispheres of Pluto. The bright and dark regions on its surface may be the result of bright frost migration.
Photo credit: A. Stern and M. Buie, NASA/ESA.

Pluto, was once in independent orbit around the Sun before it was captured, and that the two objects may have similar properties. It should be remembered, however, that Triton has probably undergone global melting as a result of its capture, and may have a much younger surface. At the time of writing, a NASA spacecraft named New Horizons is scheduled to launch in 2006, and fly by the distant Pluto and Charon system in 2015, going on to visit one or more additional Kuiper Belt objects by 2026.


## 9.7   ENDNOTE

Although it may be a long time before any spacecraft returns to the Jovian system to study the Galilean satellites, icy satellite research is far from static. Studies are being carried out in laboratories here on Earth to better understand the properties of ice at very low temperatures and over a range of pressures, and to identify the composition and mixtures of substances measured spectrally on icy satellite surfaces. The Cassini spacecraft is poised to thoroughly study Saturn and its icy moons, and to drop a probe into Titan's atmosphere, while the New Horizons spacecraft is being built for its mission to Pluto and the Kuiper Belt. In addition, there is a considerable amount of existing icy satellite data from Voyager and Galileo yet to be fully analyzed, as well as increasingly sophisticated Earth-based telescopic observations that will continue to reveal secrets about these icy worlds and their fascinating cryovolcanic processes.


## 9.8   REFERENCES

Carr, M.H., Belton, M.J.S., Bender, K., Breneman, H., Greeley, R., Head, J.W., Klaasen, K., McEwen, A.S., Moore, J.M., Murchie, S. *et al.* (1995) The Galileo Imaging Team plan for observing the satellites of Jupiter. *J. Geophys. Res.*, **100**, 18935–19956.

Cruikshank, D. (1999) Triton, Pluto and Charon. In: J. Kelly Beatty, C. Collins Peterson, and A. Chaikin (eds), *The New Solar System*. Cambridge University Press, Cambridge, U.K., pp. 285–296.

Fagents S.A., Greeley, R., Sullivan, R.J., Pappalardo, R.T. and Prockter, L.M. (2000) Cryomagmatic mechanisms for the formation of Rhadamanthys Linea, triple band margins, and other low-albedo features on Europa. *Icarus*, **144**, 54–88.

Greeley, R., Chyba, C.F., Head III, J.W., McCord, T.B., McKinnon, W.B., Pappalardo, R.T., and Figueredo, P. (2004) Geology of Europa. In: *Jupiter: The Planet, Satellites and Magnetosphere*. Cambridge University Press, Cambridge, U.K.

Harland, D.M. (2000) *Jupiter Odyssey: The Story of NASA's Galileo Mission*. Springer–Praxis, Chichester, U.K.

Kirk, R.L., Soderblom, L.A., Brown, R.H., Keiffer, S.W., and Kargel, J.S. (1995) Triton's plumes: Discovery, characteristics, and models. In: D.P. Cruikshank (ed.), *Neptune and Triton*. University of Arizona Press, AZ, pp. 949–989.

Kivelson, M.G., Khurana, K.K., Russell, C.T., Volwerk, M., Walker, R.J., and Zimmer, C. (2000) Galileo magnetometer measurements: A stronger case for a subsurface ocean at Europa. *Science*, **289**, 1340–1343.

McKinnon, W.B. (1999) Midsize icy satellites. In: J. Kelly Beatty, C. Collins Peterson, and A. Chaikin (eds), *The New Solar System*. Cambridge University Press, Cambridge, U.K., pp. 297–310.

Moore, J.M., Chapman, C.R., Bierhaus, E.B., Greeley, R., Chuang, F.C., Klemaszewski, J., Clark, R.N., Dalton, J.B., Hibbits, C.A., Schenk, P.M. *et al.* (2004) Callisto. In: F. Bagenal (ed.) *Jupiter: The Planet, Satellites and Magnetosphere*. Cambridge University Press, Cambridge, U.K.

Pappalardo, R.T., Head, J.W., Collins, G.C., Kirk, R.L., Neukum, G., Oberst, J., Giese, B., Greeley, R., Chapman, C.R., Helfenstein, P. *et al.* (1998) Grooved terrain on Ganymede: First results from high-resolution imaging. *Icarus*, **135**, 276–302.

Pappalardo, R.T., Collins, G.C., Head III, J.W., Helfenstein, P., McCord, T.B., Moore, J.M., Prockter, L.M., Schenk, P.M., and Spencer, J.R. (2004) Geology of Ganymede. In: F. Bagenal (ed.), *Jupiter: The Planet, Satellites and Magnetosphere*. Cambridge University Press, Cambridge, U.K.

Schenk P.M., McKinnon, W.B., Gwynn, D., and Moore, J.M. (2001) Flooding of Ganymede's bright terrains by low-viscosity water ice lavas. *Nature*, **410**, 57–60.

Smith, B.A., Soderblom, L.A., Beebe, R., Boyce, J., Briggs, G., Carr, M., Collins, S.A., Cook II, A.F., Danielson, G.E., Davies, M.E., *et al.* (1979) The Galilean satellites and Jupiter: Voyager 2 imaging science results. *Science*, **206**, 927–950.

Smith, B.A., Soderblom, L.A., Batson, R., Bridges, P., Inge, J., Masursky, H., Shoemaker, E., Beebe, R., Boyce, J., Briggs, G., *et al.* (1982) A new look at the Saturn system: Voyager 2 Images. *Science*, **215**, 504–536.

Smith, B.A., Soderblom, L.A., Beebe, R., Bliss, D., Boyce, J.M., Brahic, A., Briggs, G.A., Brown, R.H., Collins, S.A., Cook, A.F., *et al.* (1986) Voyager 2 in the Uranian system: Imaging science results. *Science*, **233**, 43–64.

Stern, S.A. and McKinnon, W.B. (2002) Triton's age and impactor population revisited in the light of Kuiper Belt fluxes: Evidence for small Kuiper Belt objects and recent geological activity. *The Astronomical Journal*, **119**, 945–952.

Zahnle, K., Schenk, P., Levison, H., and Dones, L. (2003) Cratering rates in the outer Solar System. *Icarus*, **163**, 263–289.

# 10

# Products of powerful volcanic explosions on Earth and Mars

*Mary G. Chapman* (US Geological Survey)
*Gudrun Larsen* (University of Iceland)

*I (Mary) have been a Research Geologist with the Astrogeology Team of the US Geological Survey (USGS) in Flagstaff, Arizona, for over 20 years. My expertise includes planetary geology, outflow channels and volcanic deposits on Mars, sub-ice volcanism, and the sedimentology of volcaniclastic deposits on the Colorado Plateau. Planetary geology involving the study of Mars is probably a very esoteric career choice to most people. People choose careers based on the serendipity of life's events, recommendations by people, and personal interest. For example, in my late teens, I enjoyed hiking and was trying to pursue a degree in science that would allow me to work outdoors. I narrowed my interest down to geology, when the subject was brought to life in an undergraduate class at the University of Utah taught by Dr. William Lee Stokes (a stratigrapher and paleontologist) and in my job as a fledgling geologist and field assistant in the Coal and Uranium Department at Utah Power & Light Company. After earning my BSc degree, I went to work as an Engineering Project Geologist with the Bureau of Reclamation in Duchesne, Utah, in charge of a pre-construction program to determine geologic suitability of the first roller-crete dam (a cement structure with a hollow core filled by compacted, in situ excavated materials) in the U.S.A. (the Upper Stillwater Dam in the Uinta Mountains). After a couple of nearly fatal accidents inside a 7'-diameter tunnel one night, I decided to earn a higher degree that would land me a safer job, so I took a better paying position as a contract Exploration Geologist with the Coal Department of Getty Oil, to save up enough money to bankroll a MSc degree. After the last drill hole was finished, I was off to the University of Northern Arizona in Flagstaff, to pursue a Master's degree in Geology and my eventual job, at the USGS. Baerbel Lucchitta, one of the first and most famous female planetary geologists, recommended me to be hired in 1983 by the USGS Astrogeology Team. In 1985, I began working for Hal Masursky and Dave Scott interpreting the Viking satellite images to produce geologic maps of Mars (the famous 1:500,000-scale quadrangles in use by planetary geologists for almost two decades). My early scientific contributions were the documentation and interpretation of catastrophic flood or outflow channels on Mars and a study of the Middle Jurassic Carmel Formation in southern Utah. The red beds of the Carmel formation rocks are a mixture of depositional environments and rock types, containing salt deposits, wind derived sand, stream channel materials, debris flow deposits, and volcaniclastic (sediments derived from volcanic materials) rocks. This field research prepared me for the varied volcanic and volcaniclastic red rocks found on Mars.*

*I (Gudrun) received my geology degree from the University of Iceland. My interest in geology came mostly from my grandfather and father, who instilled in me a respect for the environment and a curiosity concerning the nature of the surrounding rocky terrain of my native country, Iceland. Courses in Geology were not taught at the University of Iceland when I first came there in 1964. I had decided to get a BA degree and become a teacher, so I took courses in English and History in the first year. During my second year I met my future husband, Adalsteinn Eiriksson, a theology student and a teacher, and within a year we were a family of three with my teaching career postponed for the time being. The year of 1968 was a significant year in many respects (also for me). In 1968 courses in Geology appeared on the curriculum of the University of Iceland. A book on Hekla volcano by Thorarinsson was published in 1968 and a Geology textbook in Icelandic by Einarsson was also published this same year. I read these books with great interest. In the spring of 1970 the Hekla volcano erupted, and I decided that this was it—I wanted to be a geology teacher. I enrolled at the University of Iceland in the fall of 1970. One of my Geology Professors was Sigurdur Thorarinsson, internationally recognized for developing tephrachronology (the application of tephra (volcanic ash) layers as dating tool). He referred me to Dr. Sigurdur Steinthorsson, who hired me to do grain size analyses of tephra from the volcanoes Hekla and Katla. I did my BSc thesis in 1975 on a ca. 4,300 year old tephra layer we call Hekla-4, later publishing the results together with Thorarinsson (Larsen and Thorarinsson, 1977). My thesis was on the stratigraphy (systematic study of rock sedimentary sequences) of tephra around the Katla volcano and this involved summer time fieldwork, visiting remote areas, and digging holes and logging sections through through tephra and soil—I really liked this, I brought the family along and we camped in the evenings in the middle of nowhere. By 1976 research work on tephra had become my main interest and I got a stipendium at the Nordic Volcanological Institute in Reykjavik, and eventually became a research scientist at the Science Institute, University of Iceland. I currently work as a researcher at the Science Institute of the University of Iceland, specializing in the application of tephrochronology to the study of the volcanic history of Iceland. My studies of wide-spread tephra layers in Icelandic soils have made fundamental contributions to the understanding of the history of Iceland.*

Data from instruments on the orbiting Mars Global Surveyor (MGS) spacecraft are inspiring new hypotheses about the planet's surface. In particular, high-resolution (2–5 m/pixel) images from the Mars Orbiter Camera (MOC) have generated a new suggestion of possible widespread, layered sediments, some as thick as 4 km, at nearly equatorial latitudes (Malin and Edgett, 2000). Sediments are fragments of existing rock that have been transported and deposited by air, water, ice, or volcanic explosions. One of the most extensive outcrops of this material occurs in the highland area of Xanthe Terra and directly east in northern Meridiani Planum (Figure 10.1). These deposits were originally observed in older, lower resolution Mariner and Viking orbiter images, where they appear to have fewer impact craters than adjacent ancient highland cratered material. The unit was observed to bury large craters, have an intermediate albedo (relative brightness), and a friable (unconsolidated and easily eroded) surface appearance. The material was named the subdued cratered unit and was interpreted to overlie ancient highland material (Scott and Tanaka, 1986; Greeley and Guest, 1987) (see Figure 10.1).

The rocks on Mars are broken into 3 age systems, from oldest to youngest: Noachian, Hesperian, or Amazonian, based on the number and size of impact craters (see Chapter 6) on the surface. The subdued cratered unit was interpreted to be of Late Noachian age (Scott and Tanaka, 1986; Greeley and Guest, 1987; Witbeck *et al.*, 1991; Rotto and Tanaka, 1995). This material and all other units of Hesperian and Noachian age adjacent to the Valles Marineris chasmata (depressions or troughs) were modified during the formation of this canyon system, and later eroded by the outflow channels to the east (Figure 10.1(b)) (Scott and Tanaka, 1986; Witbeck *et al.*, 1991; Rotto and Tanaka, 1995). Therefore, we know the subdued crater unit formed sometime after the end of the heavy impact crater bombardment and before the formation of Valles Marineris and the outflow channels.

The subdued crater unit is interesting to planetary geologists for several reasons. First of all, composition data indicate a concentration of crystalline hematite within the subdued cratered unit in the north Meridiani Planum (Christiansen *et al.*, 1998) and in Xanthe Terra, within the dark materials of Aram Chaos (Figure 10.1(b)) (Christensen *et al.*, 2000a; Noreen *et al.*, 2000) and Valles Marineris (Noreen *et al.*, 2000; Christensen *et al.*, 2001). The hematite deposits are the only distinct, localized material that has been detected on Mars. On Earth, palagonite is a hematite-rich clay deposit that occurs in volcanic terrains and can sometimes be formed by bacteria (Thorseth *et al.*, 1992). It is possible that the Mars hematite material may have been formed by bacteria, like these terrestrial palagonite deposits. It has also been suggested that Mars hematite may have been formed by bacteria in a manner similar to the banded iron formations on Earth (Allen *et al.*, 2001). Alternatively, the hematite may have formed by hydrothermal alteration or chemical precipitation (Christensen *et al.*, 1998; Christensen *et al.*, 2000a; Christensen *et al.*, 2000b). Scattered mounds and buttes, exposed in outcrops in Meridiani Planum, are similar to terrestrial fumarolic mounds (cemented areas formed by vapor escape from cooling volcanic deposits) and provide supporting evidence of hydrothermal alteration (Chapman, 1999; Chapman and Tanaka, 2002). Possible hydrothermal areas and hematite-rich zones suggest that the Mars' sites may have exobiologic significance.

Another reason the subdued cratered unit is interesting is because MOC shows it to have numerous layers of beds that range in thickness from relatively thin to

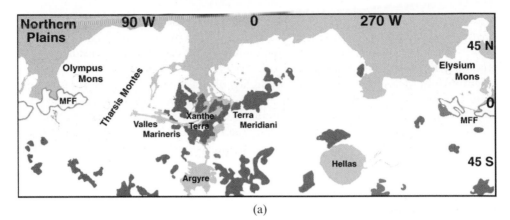

(a)

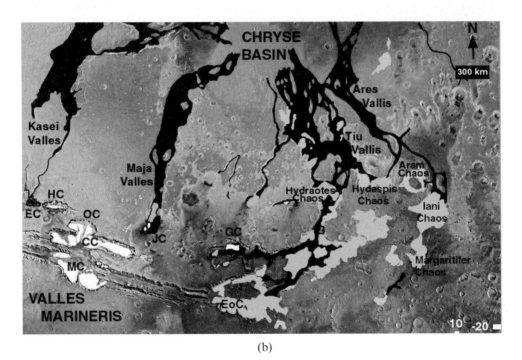

(b)

**Figure 10.1.** (a) Mars schematic map showing plotted locations of rock outcrops of the Medusae Fossae Formation (light with dark outline; labeled MFF), suggested localities of water/ice (medium gray), the subdued, layered, massive, and thin (LMT) rock unit (dark gray), and selected feature names. (b) Location map showing Valles Marineris troughs or chasma (EC = Echus Chasma, HC = Hebes Chasma, OC = Ophir Chasma, CC = Candor Chasma, MC = Melas Chasma, JC = Juventae Chasma, GC = Gangis Chasma, and EoC = Eos Chasma), and Chryse Basin; outflow channels shown in black, interior deposits shown in white, and chaos (or areas of collapse with jumbles of large blocks) are shown in gray.

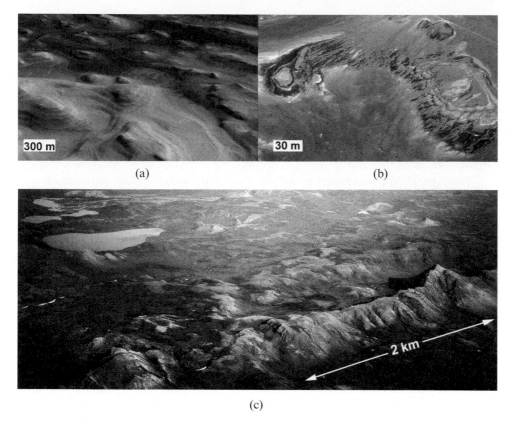

(a)                                        (b)

(c)

**Figure 10.2.** (a) Oblique view of layers in LMT deposits of a Martian crater centered at 8°W, 7°N. (b) Aerial photograph showing layered Vatnaöldur and Veidivötn ash deposits. (c) Aerial photograph of part of the Veidivötn volcanic system showing a complex terrain that consists of a volcanic fissure in the foreground and scoria cones with internal lakes in the upper left corner.

(a) Image 18-01349; courtesy of Malin Space Science Systems; (b) taken by M. Chapman; (c) Taken by M. Chapman.

massive (Figure 10.2(a)). Widespread outcrops showing layered bedding led Malin and Edgett (2000) to suggest that the subdued unit may be lacustrine or lake deposits. These authors also coined a new name for the subdued cratered unit: layered, massive, and thin (LMT) material. A lacustrine origin implies huge ancient lake volumes and would greatly alter our current understanding of the hydrogeologic history of Mars. In addition, because standing bodies of water also provide other important sites for exobiological evolution, this interpretation of lacustrine origin may have a bearing on the studies of Martian terrain and, therefore, targeting of future missions. Alternatively, these LMT materials have also been suggested to be widespread "tephra" or volcanic ash deposits (Chapman and Tanaka, 2002; Hynek et al., 2002). A tephra origin is interesting because, other than the Medusae Fossae Formation (MFF) (Sakimoto et al., 1999; see also Chapter 6), such widespread thick ash deposits have never been detected on Mars. However, the lower atmospheric pressure on Mars should yield higher eruption velocities and lower final densities of decompressed gas, which would cause eruption columns to rise much higher, have larger near-vent particles, and distribute fine particles much farther than those on the Earth (Wilson and Head, 1983; Wilson and Heslop, 1990; Wilson and Head, 1994). In addition, volcanic ash flow formation should be more likely on Mars than Earth because the reduced atmospheric pressure enhances the collapse of eruption columns (Wilson and Head, 1994). If the LMT deposits are ash, then perhaps MGS data is finally producing more definitive evidence of the widespread tephra deposits that were predicted to be likely in the reduced atmospheric pressure of Mars.

The layered Martian hematite units are visually similar to both terrestrial lacustrine and ash deposits. Both types of deposits can have variable brightness, are fine-grained, layered, massive, or thin, and can erode with gullied (or badland) topography similar to the Martian material. So in order to test the hypotheses of lacustrine versus volcanic ash origin for the hematite/LMT deposits on Mars, it seemed logical to visit these types of deposits on Earth and study the terrestrial rocks in order to determine characteristics that: (1) provide evidence for their origin; and (2) can be detected by instruments on surface rovers and spacecraft. Investigations of terrestrial rocks that appear similar to deposits on the surfaces of other planets are called analog studies.

Planetary geologists commonly perform terrestrial analog studies in order to better understand the geology of extraterrestrial worlds. Researchers in planetary geology tend to concentrate on studies that spark their personal interest and on topics on which they can contribute from their field of expertise. The lead author's affinities with volcaniclastic rocks and interests in catastrophic flood channels have led to other scientific avenues: the investigation of possible volcano/ice interactions on Mars and Earth, sub-ice volcanism, and research studies in Iceland (USGS, 2004). Because volcanism and water/ice interaction produce layered ash deposits on Earth (Chapman and Tanaka, 2002), formulating ways to investigate the origin of the layered concentrated crystalline hematite deposits on Mars peaked her interest in this analog study.

## 10.1    RATIONALE FOR ANALOG STUDIES AND HUMAN IMPETUS

Future research will likely determine the origin of the hematite/LMT units. Currently, the Mars Global Surveyor spacecraft and its on-board instruments continue to orbit the planet and acquire data, and in January 2004, NASA placed a lander on Mars equipped with a Mars Exploration Rover (MER) on the Sinus Meridiani hematite/LMT site. The hematite at the site has been found to be concentrated within round particles the size of BBs (spherules). The objective of the MER missions is to learn more about the ancient climate and putative water-rich environments through direct rock examination. The examination instruments include PanCam, Rock Abrasion Tool, 3 spectrometers (Miniature Thermal Emission Spectrometer, Mössbauer, and Alpha Proton X-Ray or APXS), and a

Microscopic Imager (Squyres, 1998; see Chapter 7). Future rover missions are also being planned for 2005, 2007, and 2009 launch opportunities. Within the next 20 years, perhaps astronauts may be sent to Mars. The orbital and future rover missions may provide evidence to determine if the LMT units are lacustrine or tephra deposits. For example, interpretation of additional TES, mini-TES, and Alpha Proton x-ray data may be able to better discriminate volcanic compositions and glasses, and clay alteration products of volcanism (such as palagonite, montmorillonite, bentonites, sideromelane, tachylite, chlorite, etc). Thermal Emission Spectrometry can view rocks in the infrared and is used to determine types and amounts of minerals: a mini-TES is a small version of this instrument that can be mounted on a rover. The Mössbauer spectrometer mounted on a rover arm can be placed against rock and soil targets to identify minerals that contain iron. The APXS instrument consists of alpha particle sources and detectors for back-scattered alpha particles, protons and X-Rays and can determine elemental chemisty of surface materials for most major elements except hydrogen. This instrument does this by exposing a material to a radioactive source that produces alpha particles with a known energy, and to acquire energy spectra of the alpha particles, protons and X-Rays returned from the sample. The PanCam is a high-resolution camera that can take panoramic images that show surface boulders/rocks with fiamme (squashed ash fragment) and vesicular (gas bubble cavity) textures. There is also the possibility that this camera could view gullies eroded into rock-rich tuffs or tuffs with ballistically emplaced (volcanically hurled) fragments that distort impacted layers of ash. Finally, the Microscopic Imager has the capability to reveal ash shards within rocks broken by the Rock Abrasion Tool.

    Missions to Mars are expensive and it is necessary to be certain our Mars rovers can detect the origin of layered sedimentary terrains like the LMT units before they are deployed on the planet. Therefore, it is wise to develop criteria to better identify the nature of these layered deposits from surface measurements and orbiting spacecraft data. For a terrestrial analog study, fluvio-lacustrine and volcanic terrains would have to be chosen that are visually similar to the LMT material outcrops, having layered and gullied deposits, and relatively free of vegetation, to simulate the barren surface of Mars. Each analog site would be geologically evaluated using sedimentary structures and rock identification to interpret the environment and processes of origin. A plan had to be formulated to identically evaluate each site to: (1) establish criteria that might be identified from space; (2) simulate visual data return of site characteristics; (3) simulate site soil characteristics of the rover wheel and trench tool disruption pattern; and (4) to determine rock compositions. Planning this type of study requires a firm understanding of the principles and processes provided by having learned geology.

    Why do people learn geology and pursue the advanced degrees to plan such research studies? That is a hard question to answer and very subjective. Each of the authors of this chapter chose geologic career paths that could merge family and research, and both chose to pursue advanced degrees at Universities in the U.K. at later stages in their lives. However, both authors followed very different paths to arrive at this joint research effort. Chapman's expertise includes planetary geology, outflow channels and volcanic deposits on Mars, Icelandic Mars analog environments, and sedimentology of volcaniclastic deposits on the Colorado Plateau (USGS, 2004). Larson's expertise is tephrachronology, specifically the study of varied ages and types of ash deposits in Iceland (Larsen and Thorarinsson, 1977). The two diverse backgrounds indicate that although each of the authors arrived at their interest in different ways, both eventually pursued academic interests that developed the skills and educational background to initiate, plan, and participate in the analog study.

## 10.2   STUDY DESIGN

For the analog study, four terrestrial layered terrains were chosen that were visually similar to many LMT material outcrops, having layered and gullied deposits: two fluvio-lacustrine and two volcanic locales. The fluvio-lacustrine areas are in Arizona, chosen because of their known characteristics, sparse desert flora, and proximity to the USGS in Flagstaff. The volcanic ash sites are located in Iceland and were chosen by Chapman because they had a high degree of visual similarity to Mars and because they were in close proximity to her related sub-ice volcanism and catastrophic flood study areas and could be easily reached. Mars' surface materials are generally much older than those of Earth. However, compared to Earth, erosion has hardly affected Mars. Similar to Mars, the Icelandic ash deposits are vegetation-free and relatively young, therefore, not greatly eroded (Figure 10.2(b)). For logistical reasons, the first analog site chosen to be studied in the summer of 2002 was the Icelandic Vatnaöldur and Veidivötn ash deposits locale. This chapter discusses the design of the comparative LMT analog studies and details characteristics of the Vatnaöldur and Veidivötn ash site.

On Earth, the Icelandic Vatnaöldur and Veidivötn ash deposits lie within the Eastern Volcanic Zone of Iceland, at the margin of the Veidivötn volcanic system. These ash deposits were erupted from the Veidivötn fissure and are products of powerful explosions bearing a strong physical resemblance to layered equatorial deposits on Mars. Unlike Earth, Mars does not show much evidence of crustal plates that move about the surface (plate tectonics). The stationary crust of Mars (see Chapter 6) allows extremely large volcanoes and volcanic fields to form, as the surface has not moved relative to the hot mantle plumes that drive magma to the surface from below. The lack of plate tectonics, higher eruptive flux rates, and voluminous, widespread volcanism on Mars should have formed numerous, thick ash layers. The Icelandic layered ash deposits are not as thick as those suggested to occur on Mars (Figure 10.2), but the study of Icelandic ash deposits can extend our knowledge of planetary comparisons as their characteristics can be used to discriminate between the proposed possible origins of the Martian layered hematite deposits.

For comparison, all 4 analog sites would have a simulated landing site and traverse within a $30 \times 30$-m grid. To establish criteria that might be identified from space, we would need to obtain local topographic maps and aerial photography at lower "Viking-like" resolution of around 16–48 m/pixel (purchased from aerial photography vendors) and at high-scale "MGS-like" resolution around 1.5 m/pixel (air photos from personal overflights) for each site. Our chosen grid for each would be located on these topographic and image bases. A survey tripod station would be set up on the corner of this grid and a GPS reading here would help to accurately locate the site.

To simulate visual data return of the site characteristics, we would establish for each 30 × 30-m grid a "real time" basic localized topographic reference map using surveying equipment and taking readings on 3–5-m intervals. Digital images of the site characteristics and rock types could later be draped over a topographic digital elevation model (DEM) to create simulated lander and rover 3-D PanCam views.

To simulate site soil characteristics of the rover wheel and trench tool disruption pattern, at each site the surface roughness would be measured using a manual hand-held tool, as well as geophysical instruments (such as a penetrometer). These instruments could provide information about soil properties such as bearing strength, penetration resistance, and porosity. Initially, this sounded like a good idea, however, it was rapidly learned that these types of measurements were impossible to use on recent, unconsolidated ash. Without comparative data from all sites, the soil characteristic tests were scrubbed.

To determine terrain compositions that simulate rover and return-sample analysis, samples would be taken at each site along a simulated "rover traverse". These samples would be examined by petrographic (microscopic examination of thin sections), SEM (scanning electron microscope), XRD (X-ray diffractometer), spectral, and chemical means, and sieve analysis would be obtained. A set of samples would be sieved for size analysis. Samples would be sent away to be made into probe-perfect uncovered thin sections (slabs of rock mounted on small glass slides that are so thin and highly polished that they can be viewed using a microscope, and without the glass cover slips the exposed minerals can be also probed by SEM methods). Petrographic analyses and point counts of the thin sections (TS), as well as X,Y mapping of TS minerals would be conducted for truth testing. SEM chemical and visual identification would be used. For organic molecule detection of carbonate and hydrogen, a lab would analyze a portion of the samples. Finally, samples would undergo spectral analysis in the TES lab at Arizona State University.

A team of collaborators had to be assembled for various aspects of the project. For the Icelandic analog sites, the initial team consisted of Mary Chapman and Andrew Russell. Geographer Andrew Russell (Mary's PhD Advisor from Keele University, UK) is an expert on the volcanically-induced catastrophic flood channels in Iceland (Russell, 2002) and had access to surveying equipment to assist the analog study. However, to expedite the geological site evaluation, each site required additional experts in the geological history of the site deposit. When geophysicist Magnus Gudmundsson (University of Iceland) heard of the plan to study the ash from the Veidivötn fissure deposits, he suggested that Mary contact Gudrun Larsen, co-author of this chapter. As an Icelandic volcanologist, Gudrun Larsen has participated in the mapping of the Veidivötn fissures swarm, and in particular studied its explosive basaltic eruptions, Vatnaöldur and Veidivötn, and their tephra deposits (Larsen, 1984). In addition to Gudrun, Thor Thordarson was invited to participate, as currently Gudrun and Thor are working together on another large eruption, the 10th century Eldgjá eruption in southern Iceland, and it was thought that their mutual ideas could be applied to the analog study. Thor Thordarson (volcanologist at SOEST, University of Hawaii and the Science Institute, University of Iceland) has specialized in flood basalt volcanism, including the 18th century Laki eruption in Iceland (Thordarson and Self, 1993).

Gudrun's expertise in the Veidivötn area originally began when she was assigned to study the area by the Nordic Volcanological Institute in Reykjavik. The basalt volcanism in the Veidivötn area appeared to be alternatively effusive and explosive without any obvious reason, according to the accepted volcanic history of the area at that time. Isotopic study of basalts from an eruption in northern Iceland had indicated that meteoric water might have been taken up by the magma—but was this the case at Veidivötn? This entire area is a mess of tephra rings, scoria cones, spatter cones, tephra flats, alluvial flats (alluvium or stream deposits even appeared in the airfall tephra from volcanic explosions), lava flows and almost without the organic soil where tephra layers from the Hekla and Katla volcanoes could have been preserved. On her initial trip to study the Veidivötn area, Gudrun climbed to the top of a small mountain to view the mess (Figure 10.2(c))—and sat down crying over her fate! Then she vowed to sort out the mess—and did! (Gudrun is currently writing her PhD thesis on the Veidivötn site at the University of Edinburgh, Scotland.)

## 10.3   THE VATNAÖLDUR/VEIDIVÖTN ASH ANALOG SITE ON EARTH

The study site lies within the Eastern Volcanic Zone of Iceland, at the margin of the Veidivötn volcanic system (Figure 10.3), which is among the most productive of volcanic areas in Iceland. In its simplest form, a volcanic system consists of a fissure swarm, which is a lane a few kilometers wide and tens of kilometers long with a number of volcanic fissures of different ages (volcanic eruptions take place on cracks which we call volcanic fissures), and a central volcano where volcanic eruptions are more frequent than on the fissure swarm, and where a caldera (basin-shaped volcanic depression) may have formed over a collapsed magma chamber. The analog site was located on tephra flats on the south-western part of the Veidivötn fissure swarm (Figure 10.3).

Veidivötn literally means "Fishing lakes". Today there are numerous lakes within this particular volcanic area and the lakes are popular for trout fishing. The camping ground and fishing huts at Veidivötn lie within one of the wide tephra rings on a 67 km long volcanic fissure that was active in the late 15th century, most likely in the year 1477 AD. Around the lakes is lush vegetation, in stark contrast to the barren, brownish flats of basaltic pumice that surround the area. Some 10,000 years ago, most of Iceland had been covered by thick ice for perhaps 100,000 years. On the south-western part of the Veidivötn fissure swarm

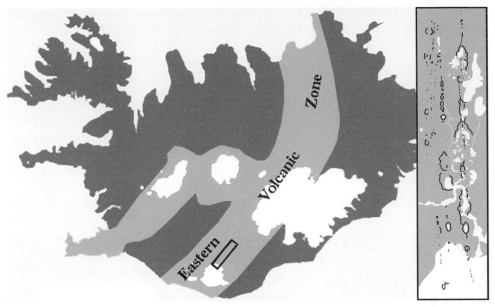

**Figure 10.3.** Schematic illustration of Iceland showing ice caps (white) and volcanic zones (medium gray); box denotes location of right inset showing crater (black) rows along Veidivötn fissure swarm.
Adapted from Larson (1984).

several volcanic eruptions had taken place below the ice, forming ridges of volcanic material called hyaloclastite (fragmental deposits of lava scattered by explosive interaction with water or ice). When the ice retreated from this area, between 10,000 and 8,000 years ago, the glacial rivers fed by the meltwater cut gorges into the bedrock and the result was a well-drained terrain within and around the volcanically active area. Like before, volcanic eruptions occurred now and then on the Veidivötn fissure swarm but now the material erupted was fluid basaltic lava, as usually happens in dry areas. The lava flowed towards the low areas and ponded or found a river gorge and was channeled down to the lowlands of southern Iceland, sometimes all the way to the seashore 100 km away, where it came to rest as a sheet of dense basaltic rock. Gradually, therefore, the lava flows raised the beds of the rivers draining the terrain until a lake area had formed within the south-western part of the Veidivötn fissure swarm, about 2,000 years ago.

Since then there have been two volcanic eruptions in this area, the Vatnaöldur eruption in the late 9th century and the Veidivötn eruption in the late 15th century (Larsen, 1984). Both of them were predominately explosive eruptions because the 1,100–1,200°C hot basaltic magma encountered water when it reached the surface and was instantaneously fragmented into sub-millimeter to centimeter-sized particles that we call tephra. Such eruptions are said to be phreatomagmatic or hydromagmatic. The tephra generally consist of three components: glass and crystals from the magma, and lithics (pieces of pre-exiting volcanic rock) from the substrata at the eruption sites. Basaltic glass may either be sideromelane, a translucent brownish glass, or tachylite, an opaque black "glass" with microscopic crystals of Fe/Ti oxide.

## 10.4   RECONNAISSANCE OF THE SITE: DAY 1

On 1 July, 2002, Andy Russell and Mary Chapman drove to the Veidivötn area and located an area for the $30 \times 30$-m grid on the aerial photos (Figure 10.4). The ash beds were very Mars-like, down to the frequent dust devils (Figure 10.5). The only difference was that the nearby cinder cones and feeder fissure swarms were too close—on Mars the vents could be so distal that we would not be able to see them from our landing site. Remember that on Mars the lower atmospheric pressure

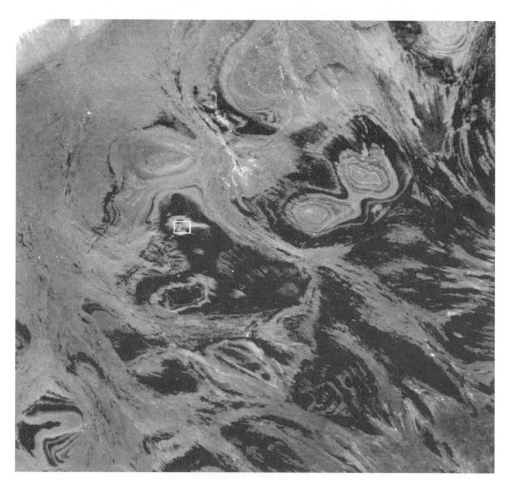

**Figure 10.4.** Commercial aerial photograph showing layered Vatnaöldur and Veidivötn ash deposits; box denotes location of $30 \times 20$-m analog study area.

**Figure 10.5.** Dust devil on Vatnaöldur and Veidivötn ash deposit; road for scale.

would distribute ash particles much farther than those on Earth (Wilson and Head, 1983; Wilson and Heslop, 1990; Wilson and Head, 1994).

The explosions during the Vatnaöldur and the Veidivötn eruptions were powerful enough to throw the tephra 10–15 km into the air and from there it was carried by winds over large areas of Iceland. Where the tephra cover was thin and did not harm the vegetation these deposits were soon covered and now appear as discrete tephra layers in the soil. Where the tephra cover was several meters thick the vegetation never recovered and the tephra remains on the surface, eroded to various degrees by wind and water. The analog study "rover traverse" is located on just such a tephra deposit, originally over 10 m in thickness.

Like the LMT deposits on Mars, the ash beds had layers that could be observed on MOC-scale aerial photos (taken a week later; Figure 10.6). These layers could be observed to form concentric bands (Figure 10.6) that on first glance looked like lacustrine shorelines (Figure 10.7). The "rover traverse" lies about 2.5 km from the Vatnaöldur volcanic fissure and the bedded material is tephra erupted at the two largest of the Vatnaöldur tephra rings. Tephra in eruptions where hot basaltic magma interacts with water is thrown out in numerous explosions and the air fall ash from each explosion may form a discrete bed in the deposit. Both the Vatnaöldur and the Veidivötn tephra deposits show distinct bedding that is still discernible up to 100 km from the source. Sometimes the eruption column may collapse, creating a surge that rolls over the rims of the craters and transports the tephra horizontally

**Figure 10.6.** Aerial photograph showing concentric layers of Vatnaöldur and Veidivötn ash deposits; box denotes 30 × 30-m analog study area.
M. Chapman.

**Figure 10.7.** Ground photograph of concentric ash layers.

outwards depositing a layer that can extend several kilometers from the source. At the "rover traverse" most of the primary Vatnaöldur deposit appeared to be air-fall tephra. The bedding was on a scale of centimeters to tens of centimeters and some of the beds contained a high percentage of sub-millimeter grains.

On the ground, the dust devils had plucked and eroded the ash layers into miniyardangs (wind erosion features shaped like the keel of a boat; Figure 10.8). In what was deemed a topographically interesting area, a 30 × 30-m grid and internal "landing site" was staked and the corners measured by tape and compass. After that, driving winds picked up along with rain, and the group decided to head to the

**Figure 10.8.** Small yardangs or boat-shaped wind erosion features on ash beds; centimeter card for scale.

**Figure 10.9.** Lake that infills one crater of the Veidivötn fissure swarm at Veidivötn campground.


campground where they would rendezvous that evening with Gudrun Larsen and Thor Thordarson (Figure 10.9).


## 10.5   SITE STUDY: DAY 2

On 2 July, we began the topographic measurements of the $30 \times 30$-m grid. It was decided that Andrew would man the survey tripod and site and record readings from the optic reflector on the staff, placed at 5-m intervals (Figure 10.10). Mary simulated a "rover traverse" within the grid that incorporated interesting terrain, by dragging her hammer through the ash. Points along the traverse were noted, described, trenched, sampled, and photographed. Gudrun and Thor, trying to expand on Gudrun's historical study of the ash deposit, dug and sampled a deep trench and then consulted on the surface materials at the rover traverse points (Figure 10.10(a)).

The site has several different interesting aspects. What appeared to be concentric layering of possibly different materials was due to wind erosion and differential drying of layers with the same chemical composition, but slightly different grain sizes (Figure 10.11).

The tephra fragments that covered the surface were highly magnetic. This material was tachylite with microscopic crystals of Fe/Ti oxide in the form of magnetite. Although the Vatnaöldur and the Veidivötn explosive eruptions formed deposits of basaltic tephra with the same composition, each deposit is different. In the Vatnaöldur eruption a nearly 30 km long part of the volcanic fissure opened up along the edge of a lake basin where water could interact with the erupting magma and the result was highly fragmented tephra. The basaltic magma was rich in transparent colorless crystals of plagioclase, and therefore the Vatnaöldur tephra contains abundant small colorless crystal fragments together with the greenish-

(a)                                                    (b)

**Figure 10.10.** Analog study group working at the Vatnaöldur and Veidivötn site. (a) Andrew Russell at the survey station; Gudrun Larsen and Thor Thordarson in the background at the trench site. (b) Field assistant Lawrence Tanner with a survey rod.

**Figure 10.11.** Concentric layering ash formed by wind erosion and differential drying of layers with the same chemical composition, but slightly different grain sizes; light layer central to the mound is dry relative to the layers above and below; hammer for scale.

brown volcanic (sideromelane and tachylite) glass. The magnetic tephra particles formed dry dunes and pocket fills on top of, and adjacent to, wet material (Figure 10.12(a)). Initial observations suggested that the magnetic ash was a later dune-forming eolian (wind) deposit that migrated across the surface. However, these

(a)                                                        (b)

**Figure 10.12.** Magnetic tephra particles form dry dunes (a) and pocket fills (b) on top of and adjacent to wet material; hammer for scale. Magnetic particles shown intermixed with wet, finer grained ash, right of hammer butt on (b).

same magnetic particles could be found adjacent to the dry dunes, intermixed with wet, finer grained ash (Figure 10.12(b)). Therefore, it is more likely that wind had winnowed out the fine ash from the dry dune deposits. The wet material was too cohesive to allow wind winnowing. The miniyardangs observed the previous day, could be observed within the grid to be formed of fine ash, downwind of large volcanic cinders (Figure 10.13).

Randomly dispersed within the mostly gray volcanic ash were small rocks that were white to red (Figure 10.14(a)). Many of these rocks showed clear internal layering (Figure 10.14(b)). These rocks, however, were remnants of tephra dispersed over the area by the Veidivötn eruption. In the Veidivötn eruption about 30 km of the volcanic fissure opened up along the middle of the lake basin and also extended across the river delta of a large glacial river. This time the basaltic magma encountered wet lake sediments and wet gravel deposits as well as the water. The interaction with water was less effective and as a result the tephra was coarser grained (i.e., less fragmented, containing abundant small rocks from the river delta). These rocks appeared identical to layered sedimentary rocks derived from lacustrine environments, but actually they were fragments of a silica-rich volcanic rock called flow-banded rhyolite. As coincidence would have it, the "rover traverse" was placed just where the western margin of the 600 years younger Veidivötn tephra overlaps

**Figure 10.13.** Small yardangs or boat-shaped wind erosion features downwind of large tephra clasts; hammer for scale.

the Vatnaöldur tephra deposit. The tephra that forms the western margin of the Veidivötn tephra was erupted up through the river delta (some 10 km south of "rover ground") and was loaded with water-worn pebbles of rhyolite. Here the Veidivötn tephra was originally about 50 cm thick and it still exists as patches, but at the "rover traverse" it had mostly been eroded away by wind and water. Only the heaviest component, the rock material, was left as a residual deposit. A great deal of those rocks were water-worn rhyolite pebbles, ranging from < 50 cm to >3 cm in diameter.

In summary, the analog site sits atop the 1,100 year old, bedded, primary Vatnaöldur deposit of basaltic tephra, eroded by wind and water to expose the layering of the ash beds, in places overlain by patches of reworked Veidivötn and (probably) Vatnaöldur tephra, and strewn with water-worn rhyolite pebbles from a river delta 10 km away that had fallen from the air in the younger eruption some 600 years later.

Following completion of the field portion of the site study, the group went off to examine the source vents for some of the ash deposits (Figure 10.15). The two Vatnaöldur tephra rings (also referred to as "ash rings" because a large part of the tephra classifies as an ash fraction (i.e., grains are smaller than 2 mm)) were huge, the larger one being about 1.5 km wide between the rims with the crater having a lake at the bottom. We followed a track used by fishermen down to the lake and viewed the tephra ring from the inside. Near the top the bedded tephra that forms the rims could be seen, but the inner slopes were partly covered with scree and loose tephra, which made vehicle traverse quite difficult at times.

**Figure 10.14.** (a) Randomly dispersed small, round, white to red colored cobbles; hammer for scale. (b) Layered clasts are vesicle-free and fine-grained.

## 10.6   DISCUSSION

At this point in time (early 2004) the study is not complete. A digital 3-D simulated "rover and lander" view has not yet been generated. Collected samples are still

**Figure 10.15.** Part of western rim of 1.5 km diameter tephra ring source for the Vatnaöldur ash deposits; bedded tephra crops out at the top of the crater, above scree-covered slopes.

waiting for the analytical testing. Therefore, our discussion is currently limited to field insights from the Veidivötn study.

The apparent layering of the Vatnaöldur and Veidivötn ash may have implications for the origin of the LMT materials on Mars. Layered beds of varied relative brightness characterize the hematite and LMT materials. These beds lack any other geomorphologic feature that woul$ be diagnostic of their origin other than that they locally erode into streamlined knobs, which are likely yardangs or small, scattered, non-streamlined mounds or buttes (Chapman and Tanaka, 2002). The lacustrine (lake or water-laid) sedimentary interpretation of the LMT material is based on: (1) the relatively low and flat local topography of some areas; (2) possible salt deposits formed by evaporation of standing bodies of water (evaporites) noted by Lee (1993) within large, old craters; (3) dunes and Viking data that indicate sand-sized material; (4) few channels (valley networks) produced by running water; (5) the smooth surface of the hematite material; and (6) numerous outcrops that show horizontal layering, and concentric bands of layers (Edgett and Parker, 1997; Malin and Edgett, 2000). However, topography from the Mars Orbiter Laser Altimeter (MOLA; on board the MGS spacecraft) indicates that the Meridiani Planum hematite deposit occurs on a gentle slope—an unlikely position for lake deposits (Chapman, 2000). Alternatively, volcanic eruptions generate ash flow falls that can blanket slopes. Although sand-sized and larger material obviously can be found in water-laid sediments (Edgett and Parker, 1997), these grain sizes are also common to volcanic ash flows (Murai, 1961; Fisher and Schmincke, 1984). Tephra deposits would be expected to lack valley networks, as they do not require running wate2. Contrary to Viking observations, MOC images show that the unit is not smooth, but contains numerous fine-scale ridges (Edgett and Malin, 2000). Finally, our field study shows that volcanic ash deposits can have nearly horizontal layers that vary in relative brightness and show concentric layering. Very little evidence exists for the widespread, deep, equatorial, highland lakes or seas on Mars. Typical rock compositions of lakes have not been indicated: spacecraft instruments have not shown carbonates (sediments formed by precipitation of calcite) or evaporites. Case in point: "White Rock", another famous suspected Mars crater-fill evaporite, lacks spectral evidence for a water-laid origin (Ruff *et al.*, 2000). However, volcanic rocks occur abundantly on Mars and the reduced atmospheric pressure of Mars should produce widespread ash deposits. Our field study suggests ruling out horizontal and concentric bands of layers as a definitive characteristic of lacustrine origin and keeps open the possibility of a volcanic origin for the LMT deposits. In addition, site studies indicate that apparent layers may be created by variable drying of ash of identical composition, but of different particle size distribution.

The banded rhyolite clasts at this site indicate that volcanic eruptions can introduce rounded volcanic rocks from distal locales that show internal layering, similar to sedimentary rocks. The composition of these rocks could not be deduced by hand examination in the field or image data from Mars. Compositional analysis would be required.

The highly magnetic tachylite cinders at the analog site have implication for the concentrated crystalline hematite site on Mars. Christensen *et al.* (2000a, 2000b) have considered five possible mechanisms for the formation of the crystalline hematite: (1) chemical precipitation that includes origins by: (a) precipitation from standing, oxygenated, Fe-rich water, (b) precipitation from Fe-rich hydrothermal fluids, (c) low-temperature dissolution and precipitation through mobile groundwater leaching, and (d) formation of surface coatings; and (2) thermal oxidation of magnetite-rich lava. They favor chemical precipitation models involving precipitation from Fe-rich water based on: (1) the perceived association of the hematite with lacustrine sedimentary materials (Edgett and Parker, 1997); (2) the large geographic area in which hematite has been detected; (3) an apparent large distance of the hematite from regional heat sources; and (4) the lack of evidence for extensive hydrothermal groundwater processes elsewhere on Mars (Christensen *et al.*, 2000a). However, crystalline hematite can occur from thermal oxidation of magnetite or other Fe oxides common in terrestrial tachylitic ash deposits, such as those in the Iceland field study site, which on Mars should be widespread across a large geographic area with variable slopes and elevations. The LMT material is a plains unit that, south of Chryse Basin, is associated with chaos (areas of collapse with chaotic jumbles of large blocks) and chasmata (large troughs; Figure 10.1(b)). Chaos and chasmata have been interpreted to be huge, possible eruptive sites that could have vented widespread ash flows and provided regional heat sources (Chapman, 2000; Chapman and Tanaka, 2002).

Additional spectral studies suggest that the hematite may be composed of axis-oriented (short axes of the crystals oriented up and the long axes parallel to the ground) hematite grains, which may occur as metamorphosed (changed in form due to pressure and/or heat), schistose (flattened and bent crystalline texture) hematite deposits or as very well aligned, platy, discrete hematite particles (Lane et al., 2000). For the hematite to be schistose and originally water-laid, the fine-grained, red hematite would need to be recrystallized to coarser, gray hematite through a low-grade metamorphism produced by burial by as much as 4 km of material (Lane et al., 2000). This overburden would then have to be stripped away, leaving the schistose hematite exposed at the surface of Mars. However, this thickness of putative lacustrine overburden would be approximately 4 km higher in elevation than the shoreline contact envisioned by Edgett and Parker (1997) at Meridiani Planum and lake sediments can not be deposited above the water line (Chapman and Tanaka, 2002).

To form Mars-like hematite from volcanic ash requires a mechanism to align Fe oxides in tachylite, so alteration could perhaps produce aligned, platey hematite particles and cement the deposit in place. (Cementation would prevent the deposit from being eroded by the Martian winds.) The highly fragmented tephra of phreato-magmatic eruptions like Vatnöldur and Veidivötn cool relatively fast. However, tachylite fragments erupted in air, cool slower than other types of cinders formed in phreatomagmatic eruptions (like sideromelane), as water has a much higher heat capacity and conductivity than air with cooling rates being faster than when lava is erupted in air (Fisher and Schmincke, 1984). Based on our field observation of the Vatnaöldur ash, we suggest that platey, lithified (hardened or cemented) hematite grains may have occurred by thermal oxidation of Fe oxide-rich tachylitic spatter (hot lava drop) deposits. Welding of spatter could align, thermally alter magnetite grains, and produce a lithified deposit. Is it possible on Mars to have volcanic ash remain hot enough to weld after deposition? The answer is yes, but the amount and retention of heat depends more on the gas content of the magma and the mass flux than the composition (pers. commun. Lionel Wilson). A low enough volatile content means little magma fragmentation and, therefore, coarse particles big enough to retain a lot of heat. Coarse particle size combined with low ejection speeds, not much time to cool in flight, and shared heat on landing, could reheat cooled skins enough to weld and/or form flow (rheologic) features (pers. commun. Lionel Wilson). Rheologic flow is known to produce banded, flattened, and aligned shards and crystals (Chapin and Lowell, 1979; Peterson, 1979) and could account for alignment of Fe oxide grains. Mass flux is also an important factor in producing welding. For a given gas content a bigger mass flux means more volcanic rock particles in flight at any one time, shorter intervals between particle impact, and hence protection of the underlying ash layers from further cooling (Head and Wilson, 1989). However, tachylite glass fragments cool faster than other types of basaltic cinders. We have yet to determine whether this type of glass cools too fast for the Meridiani Planum deposit to be related to putative distal chaos or chasmata sources. A comparison of the topography and the hematite deposits at Meridiani Planum indicates that the highest intensity levels of the mineral are associated with the southern half of a 200 km long, 70 km wide north-eastern trending ridge (Chapman and Hare, 2002). Mapping of the surrounding area, from topography and Viking images (which includes brightness variations) discriminates numerous lineations (7rinkle ridges, topographic breaks, possible faults) that also exhibit a north-eastern trend. It is possible that the hematite ridge may be a proximal volcanic vent that utilized an ancient, Noachian structure in the area.

Alternatively, rather than welding, devitrification (or the natural time-induced conversion of unstable glass to a crystalline rock) of tachylite ash from a distal source could form an indurated (hardened or cemented) layer rich in Fe oxide. Chemical alteration of Fe oxides by exposure to water (or perhaps repetitive carbon dioxide frosts) may have converted the material to pedogenic (soil-formed) spherules of crystalline hematite. The proposed mechanisms for deriving crystalline hematite from altered ash could explain its occurrence within the LMT materials of different ages in north Meridiani Planum, Xanthe Terra, within dark materials of Aram Chaos, and Valles Marineris (Chapman, 2002). Deposition of Fe oxide-rich ash deposits could occur in different locales at different periods in time.

Our future work will contain the results from analysis and testing. In addition, as this is an ongoing comparative study between volcanic ash and lacustrine beds, we obviously have a lot more work to do in order to pinpoint definitive characteristics of environments and origins that can be identified by a spacecraft or rover.

## 10.7   CONCLUSIONS

Possible widespread, sedimentary materials, some as thick as 4 km, occur at nearly equatorial latitudes on Mars and contain local deposits of concentrated crystalline hematite. Study of the layered material is important as its origin is contentious, the hematite deposit may be related to bacterial deposition, and one site within the deposit is a prime candidate for a future landed mission on Mars.

The layered material bears a great resemblance to lacustrine deposits and volcanic ash sediments on Earth and studies of 2 lacustrine and 2 volcanic sites are underway. A field study of one of the analog sites, the Vatnaöldur and Veidivötn ash deposit in Iceland has helped to provide criteria to better identify the nature of volcanic layered deposits from surface measurements and orbiting spacecraft data. These criteria are necessary to be certain our Mars rovers can detect the origin of layered sedimentary terrains like the LMT units before they are deployed on that planet.

The Veidivötn volcanic system, within the Eastern Volcanic Zone of Iceland, is among the most productive volcanic areas in Iceland. The analog site was located on tephra flats on the south-western part of the Veidivötn fissure swarm, a lane a few kilometers wide and tens of kilometers long with a number of volcanic fissures of different ages. Some 10,000 years ago, most of Iceland had been covered by thick ice for perhaps 100,000 years. When the ice retreated from this area, between 10,000 and

8,000 years ago, volcanic eruptions on the Veidivötn fissure swarm produced fluid basaltic lava that ponded a lake area within the south-western part of the Veidivötn fissure swarm, about 2,000 years ago. The Vatnaöldur eruption in the late 9th century and the Veidivötn eruption in the late 15th century were (mostly) explosive eruptions when magma encountered water at the surface and was instantaneously fragmented into small (sub-millimeter to centimeter-sized) ash particles called tephra.

The entire analog study, which includes three other sites, is still in progress. However, initial field insights from the Veidivötn area suggest ruling out horizontal and concentric bands of layers as a definitive characteristic of lacustrine origin and keeps open the possibility of a volcanic origin for the LMT deposits. Layers within ash may not reflect separate beds or different origins widely separated in time, but can be created by subsequent explosions, changes in particle size, composition, or degrees of welding during one eruptive cycle. In addition, differential drying of beds produced in one eruption can form the appearance of layering. On Mars, vesicle-free, fine-grained, layered, volcanic rocks will require compositional analysis to determine their origin. Deposition and alteration of Mars ash similar to Fe oxide-rich tachylite tephra at the analog site could provide a mechanism to generate hematite deposits on Earth. Welding or devitrification of ash could have cemented the Martian materials in place and, therefore, provided a mechanism that preserved the hematite deposit from wind erosion. Later, soil-forming processes could leach iron and form round pedogenic spherules of hematite. Finally, obtaining field measurements on Earth and making sense of them requires time, spare tools, and skilled experience—this task on Mars will be even more difficult, especially if we rely only on robots with limited tools and learning capacities.

## 10.8   REFERENCES

Allen, C.C., Westall, F., and Schelble, R.T. (2001) Importance of a Martian hematite site for astrobiology. *Astrobiology*, **1**, 111–123.

Chapin, C.E. and Lowell, G.R. (1979) Primary and secondary flow structures in ash-flow tuffs of the Gribbles Run paleovalley, central Colorado. In: C.E. Chapin and W.E. Elston (eds), *Ash Flow Tuffs*. Geol. Soc. of Amer. Spec. Paper 180, 137–154.

Chapman, M.G. (1999) Enigmatic terrain of north Terra Meridiani, Mars. In: Abstracts of papers submitted to the 30th Lunar and Planetary Science Conference: Lunar and Planetary Institute, March 15–18, Houston, Texas, LPSC 30th CD, #1294.

Chapman, M.G. (2002) Layered, massive, and thin sediments on Mars: Possible Late Noachian to Late Amazonian tephra? In: Smellie, J.L. and Chapman, M.G. (eds), *Volcano–Ice Interactions on Earth and Mars*. Geological Society, London, Special Publications 202, 273–303.

Chapman, M.G. and Hare, T. (2002) Sinus Meridiani hematite deposits, Mars: Results of a GIS-based study of regional MGS and Viking data sets. In: Abstracts of papers submitted to the 33rd Lunar and Planetary Science Conference: Lunar and Planetary Institute, March 11–15, Houston, Texas, LPSC 33rd CD.

Chapman, M.G. and Tanaka, K.L. (2002) Related magma-ice interactions: Possible origin for chasmata, chaos, and surface materials in Xanthe, Margaritifer, and Meridiani Terrae, Mars. *Icarus*, **155**(2), 324–339.

Christensen, P.R., Anderson, D.L., Chase, S.C., Clancy, R.T., Clark, R.N., Conrath, B.J., Keiffer, H.H., Kuzmin, R.O., Malin, M.C., Pearl, J.C. *et al.* (1998) Results from the Mars Global Surveyor Thermal Emission Spectrometer. *Science*, **279**, 1692–1698.

Christensen, P.R., Malin, M., Morris, D., Bandfield, J., Lane, M., and Edgett, K. (2000a) The distribution of crystalline hematite on Mars from the Thermal Emission Spectrometer: Evidence for liquid water. 31st Lunar Planet. Sci. Conf. Abs. Lunar and Planetary Institute, March 15–18, Houston, Texas, LPSC 31th CD, #1627.

Christensen, P.R., Bandfield, J.L., Colark, R.N., Edgett, K.S., Hamilton, V.E., Hoefen, T., Kieffer, H.H., Kusmin, R.O., Lane, M.D., Malin, M.C., *et al.* (2000b) Detection of crystalline hematite mineralization on Mars by the Thermal Emission Spectrometer: Evidence for near-surface water. *J. Geophys. Res.*, **105**(E4), 9623–9642.

Christensen, P.R., Morris, R.V., Lane, M.D., Bandfield, J.L., and Malin, M.C. (2001) Global Mapping of Martian hematite mineral deposits: Remnants of water-driven processes on early Mars. *J. Geophys. Res.*, **106**, 23873–23885.

Edgett, K.S. and Malin M. (2000) The new Mars of MGS MOC: Ridged layered geologic unit (they're not dunes; abs.). 31st Lunar and Planetary Science Conference, Lunar and Planetary Institute, March 15–18, Houston, Texas, LPSC CD, #1057.

Edgett, K.S. and Parker T.J. (1997) Water on early Mars: Possible subaqueous sedimentary deposits covering ancient cratered terrain in western Arabia and Sinus Meridiani. *Geophys. Res. Lett.*, **24**, 2897–2900.

Fisher, R.V. and Schmincke, H.U. (1984) *Pyroclastic Rocks*. Springer-Verlag, New York, 472 pp.

Greeley, R. and Guest, J.E. (1987) Geologic map of the eastern equatorial region of Mars. *US. Geol. Surv. Misc. Invest. Ser. Map*, I-1802-B, 1 : 15,000,000 scale.

Head, J.W. and Wilson, L. (1989) Basaltic pyroclastic eruptions: Influence of gas-release patterns and volume fluxes on fountain structure, and the formation of cinder cones, spatter cones, rootless flows, lava ponds and lava flows. *J. Volcanol. Geotherm. Res.*, **37**, 261–271.

Hynek, B.M., Arvidson, R.E., and Phillips, R.J. (2002) Geologic setting and origin of Terra Meridiani hematite deposit on Mars. *J. Geophys. Res.*, **107**, 18,1–18,14.

Lane, M.D., Morris, R.V., and Christensen, P.R. (2000) Sinus Meridiani shows spectral evidence for oriented hematite grains. 31th Lunar Planet. Sci. Conf. Abs. Lunar and Planetary Institute, March 15-18, Houston, Texas, LPSC 31th CD, #1140.

Larsen, G. (1984) Recent volcanic history of the Veidivötn fissure swarm, southern Iceland— An approach to volcanic risk assessment. *J. Volcanol. Geotherm. Res.*, **22**, 33–58.

Larsen, G. and Thorarinsson, S. (1977) H-4 and other acid Hekla tephra layers. *Jökull*, **27**, 28–46.

Lee, P. (1993) Briny lakes on early Mars? Terrestrial intracrater playas and Martian candidates, Abstract for the Workshop on Early Mars: How warm and how wet? LPI Tech. Rept. 93-03, Lunar Planet. Inst., Houston Texas, p. 17.

Malin M.C. and Edgett, K.S. (2000) Sedimentary rocks of Mars. *Science*, **290**, 1927–1937.

Murai, I. (1961) A study of the textural characteristics of pyroclastic flow deposits in Japan. *Tokyo Univ. Earthq. Res. Inst. Bull.*, **39**, 133–248.

Noreen, E., Tanaka, K.L., and Chapman, M.G. (2000) Examination of igneous alternatives to Martian hematite using terrestrial analogs. GSA abs. with progs. 32, no. 7, A303.

Peterson, D.W. (1979) Significance of flattening of pumice fragments in ash-flow tuffs. In: C.E. Chapin and W.E. Elston (eds), *Ash Flow Tuffs*. Geol. Soc. of Amer. Spec. Paper 180, 195–204.

Rotto, S. and Tanaka K.L. (1995) Geologic/Geomorphologic map of the Chryse Planitia region of Mars. *US. Geol. Surv. Misc. Invest. Ser. Map*, I-2441, scale 1 : 5,000,000.

Ruff, S.W., Christensen, P.R., Clark, R.N., Kieffer, H.H., Malin, M.C., Bandfield, J.L., Jakosky, B.M., Lane, M.D., Mellon, M.T., and Presley, M.A. (2000) Mars "White Rock" feature lacks evidence of an aqueous origin. 31th Lunar Planet. Sci. Conf. Abs. Lunar and Planetary Institute, March 15–18, Houston, Texas, LPSC 31th CD, #1945.

Russell, A.J. (2002) The effects of glacier-outburst flood flow dynamics on ice-contact deposits: November 1996 jökulhlaup, SkeidarÃrsandur, Iceland. In: I.P. Martini, V.R. Baker, and G. Garzon (eds), *Flood and Megaflood Deposits: Recent and Ancient*. Spec. Pub. Internat. Asso. Sedimentology, 32, 67–83.

Sakimoto, S.E.H., Frey, H.V., Garvin, J.B., and Roark, J.H. (1999) Topography, roughness, layering and slope properties of the Medusae Fossae Formation from Mars Orbiter Laser Altimeter (MOLA) and Mars Orbiter Camera (MOC) data. *J. Geophys. Res.*, **104**(E10), 24141–24154.

Scott, D.H. and Tanaka, K.L. (1986) Geologic map of the western equatorial region of Mars. *US. Geol. Surv. Misc. Invest. Ser. Map*, I-1802-A, 1 : 15,000,000 scale.

Squyres, S.W. (1998) The Athena Mars Rover science payload (abs.). Mars Surveyor 2001 Landing Site Workshop. NASA-Ames Research Center, January 26–27, Mountain View, California (http://cmex.arc.nasa.gov/Mars_2001/Squyres_abs.html).

Thordarson, T. and Self, S. (1993) The Laki (Skaftár Fires) and Grímsvötn eruptions in 1783–1785. *Bull. Volcanology*, **55**, 233–263.

Thorseth, I.H., Furnes, H., and Heldal, M. (1992) The importance of microbiological activity in the alteration of natural basaltic glass. *Geochim. Cosmochim. Acta*, **56**, 845–850.

USGS (2004) http://astrogeology.usgs.gov/About/People/MaryChapman/#Selected Publications.

Wilson, L. and Head, J.W. (1983) A comparison of volcanic eruption processes on Earth, Moon, Mars, Io, and Venus. *Nature*, **302**, 663–669.

Wilson, L. and Head, J.W. (1994) Mars: Review and analysis of volcanic eruption theory and relationships to observed landforms. *Rev. Geophys.*, **32**, 221–264.

Wilson, L. and Heslop, S.E. (1990) Clast sizes in terrestrial and Martian ignimbrite lag deposits. *J. Geophys. Res.*, **95**(B11), 17309–17314.

Witbeck, N.E., Tanaka, K.L., and Scott, D.H. (1991) The geologic map of the Valles Marineris region, Mars. *US. Geol. Surv. Misc. Invest. Ser. Map*, I-2010, scale 1 : 2,000,000.

# 11

## From Yellowstone to Titan, with sidetrips to Mars, Io, Mount St. Helens and Triton

*Susan W. Kieffer* (University of Illinois)

*I am a Professor of Geology and Charles R. Walgreen, Jr., University Professor at the University of Illinois at Urbana–Champaign. My work on volcanoes seems to have evolved as a random walk through a double major in physics and mathematics at Allegheny College, and a MSc in Geological Sciences and PhD in Planetary Sciences at the California Institute of Technology. None of these studies had one single bit of volcanology in their content! I studied how meteorite impacts altered rocks, and how atoms vibrate in complicated minerals so that I could predict their thermodynamic properties. I was, however, fascinated by the attempts in the early 1970s of several leading researchers to quantify how volcanoes work, and I developed the idea described in this chapter that I could eventually study volcanoes by initially figuring out how geysers worked, while continuing with my impact and thermodynamic studies. I envisioned the geyser studies as a 5-year plan. Once I began that work in 1975, it seemed that geysers and volcanoes erupted all over the solar system. I was tugged back and forth between Old Faithful Geyser, Io, Old Faithful Geyser, Mount St. Helens, Old Faithful Geyser, Triton—always trying to get back to finish my 5-year plan to understand Old Faithful, a plan now in its 29th year! I have arranged this chapter to show how this tugging back and forth reveals one way in which science progresses, and how it led to the theoretical models that I propose for different geyser and volcanic eruptions.*

### 11.1 PROLOGUE

It is March, and I am walking the boardwalks of Yellowstone National Park at night. All lies quiet. Stars glisten brilliantly through the clear night sky of Wyoming. Here and there, thick icy fogs enshroud fumaroles where steam leaks from the underworld. Suddenly, I am startled by a quiet "whoosh" as a large geyser erupts. Minute by minute the eruption builds; I am enveloped in a cold and moist fog and a drizzle of water droplets rains down on me. Then, almost as suddenly as it started, the geyser stops. All is quiet ... briefly. I pause to enjoy the night in the geyser basin, and soon the geyser erupts again, and again ...

During my pause, I imagine that I am on Io, the fiery satellite of Jupiter. For a moment, the sulfurous, chromatic surface lies quiet. Perhaps stars glisten brilliantly through the tenuous night sky. Here and there, low-lying ice clouds enshroud fumaroles where sulfur dioxide leaks from the underworld. Suddenly, a fissure splits the surface and billowing clouds of sulfurous gases hurl ice and ash into the sky, creating a local atmosphere in the vacuum. Minute by minute, the intensity of the eruption builds and stars begin disappearing from the night sky. Particles of sulfur and sulfur dioxide snow, and ash rise to 300 km, and later rain down across the planet 1,000 km away.

As I ponder whether this is a geyser or volcanic eruption, an area the size of Alaska is rapidly engulfed by the ash and ice products of this eruption. Under the veil of plume ejecta, dense clouds of gas and ash roll across the land, their particles mixing with ash falling from the sky; lava flows rush in torrents down the slopes. Pele, the Hawaiian goddess, is playing with Loki, Marduk, and other gods and goddesses of volcanism on Io. Time and time again, they repaint the land with different sulfurous colors. Their paint brushes are geysers, volcanoes, lava flows, ash hurricanes, and ash falls; their paint pots are reservoirs of molten sulfur, sulfur dioxide, and silicate material that exist in the underground and atmosphere of Io.

The mythical figures of volcanism play and paint until the excess heat or fluid is expunged. Then, almost suddenly, the gods, goddesses, magma, and sulfur retire into the underworld, having sculpted anew the surface, refreshed the atmosphere, and cooled the tempers of the fiery interior. All is quiet ... briefly.

## 11.2   INTRODUCTION

*A "complex system" is one in which increasing the throughput of some quantity causes an unexpected event.*

A. Hubler, pers. commun. May, 2003

We are surrounded by complex systems in our regular lives, but they are so common that we do not think of them as complex systems. Think about the simple act of making popcorn. When my son was small, I decided that he should see what happens (from a safe distance) when you make popcorn without a lid on the popcorn popper. The popcorn popper contained unpopped corn and a bit of vegetable oil. The "throughput" to the system was the increasing temperature of the oil and popcorn in the popper. For a long time, nothing spectacular happened. Then, all of a sudden, we had popcorn showering over us and coating the living room floor. The "unexpected response" was the volume change of popcorn kernels into "popped corn". Now, I expected this to happen because I had seen it before, but it was truly an "unexpected" event to a 1 year old, and is an "unexpected event" in the context of complex systems analysis!

As a second example, more relevant to geysers (Figure 11.1; see color section), think about a simple whistling kettle on a stove. Depending on the elevation where you live, water boils at different temperatures because the pressure of the atmosphere is different, and this affects the boiling temperature. At sea level, it boils at 100°C; at 2,000 m (about 7,000 ft), it boils at about 92°C. If you fill the kettle with water and turn the heat on low, the water may just warm up and steam may never burst forth to blow the whistle. The geologic equivalent in a hydrothermal area is called a "hot spring"—a pool of warm water that stays at nearly a constant temperature, below the boiling point of water. This is not a complex system, because nothing unexpected happens. Now, think of the opposite case—if you turn the heat to its highest setting on your stove, your kettle will pour out steam, and blow the whistle constantly (until it runs dry and you burn the pot). While the kettle is pouring out steam, it is not a complex system. The geologic equivalent of this condition is a "fumarole"—a place on the Earth where steam, often superheated above the boiling point of water, pours continuously from a spot in the ground. (However, this example is a bit tricky: as your kettle heats up from cold water to hot steam, the throughput of energy is increasing and the system does something unexpected: it boils and turns to steam. Similarly, when you run out of water, you have decreased the throughput of water mass until the pot burns, also an unexpected event. We ignore these parts of the problem for simplification.) Systems like this can be complicated, for example, because of chemical reactions, turbulence, and buoyancy effects, but not complex in the sense of Hubler's definition above.

Between these two extremes, you can imagine that there is one specific setting of heat on the stove that will lead to occasional bursts of steam that will intermittently trigger the whistle. Each little whistle could be thought of as an "eruption". The geologic equivalent is a "geyser". The kettle "eruption" stops when all the water is boiled off or if you turn off the heat; the geyser eruption stops either when the system runs out of water or out of heat. A geyser is more complex than a kettle, but it is a good way to start to think about the components of a geyser system, and a good way to start thinking about complex systems.

A geyser is a reservoir into which hot (and sometimes cold) water flows until the system boils vigorously enough to eject water. The word "geyser" is derived from the Icelandic word "gosha", which means "to burst forth". It was also the name of the Great Geyser of Iceland, now extinct. Geysers are relatively rare and unique features because they can only exist when there is a unique balance of water and energy to heat the water to the boiling point. With few exceptions, they occur in areas where volcanic activity either is, or has been, intense, and where melted rock ("magma") around volcanoes provides heat to groundwater that arrived via the normal meteorological cycle (i.e., rain, snowmelt, and groundwater flows).

When sufficient energy flows into the water that the volume fraction of vapor exceeds about 70%, an "unexpected response" occurs: boiling liquid changes into a droplet-laden aerosol generating a huge volume change. Just like expanding popcorn being thrown out of the popper, this volume change drives the water–steam mixture

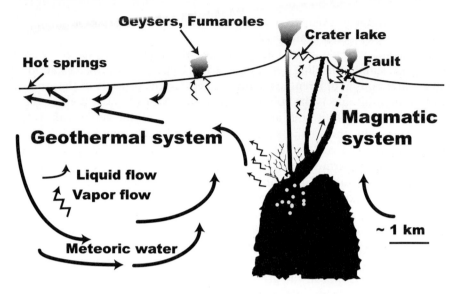

**Figure 11.3.** A schematic cross section of a volcano, with its magma reservoir(s) and circulating hot water and vapor.

from the reservoir. An eruption occurs. The "increasing throughput" that makes Old Faithful a complex system is the increasing temperature or energy of the liquid in the reservoir. The increase of temperature or energy with time leads to the "unexpected event", the eruption.

Volcanoes are much more complicated than geysers. The "increasing throughput" might differ from one volcano to another. For some volcanoes, such as those exhibiting phreatic or phreatomagmatic activity, the throughput might be the increasing amounts of water entering the system, resulting in repeated bursts of "unexpected" activity separated by interludes of quiet. This was the case in the March–April, 1980, eruptions of Mount St. Helens, in which steam-driven fragments of wall rock were carried upward in steam-driven eruptions (Figure 11.2; see color section).

However, there are other possibilities for the throughput variable in volcanism. An active volcano has magma, as well as water, within its plumbing system (Figure 11.3). In times of dormancy, the magma is stationary—resting a few kilometers underground, providing heat to nearby rocks and groundwater. The magma can cool slowly during such times without a volcanic eruption. However, the composition of the magma may change. For example, gas may exsolve from the magma as it cools, leading to an eruption. In this case, the gas exsolution rate could be the increasing throughput that leads to the unexpected—the eruption. Alternatively, magma may move higher into the volcanic edifice, either due to tectonic activity or to an injection of fresh magma at depth that forces it higher. In the case of the eruption of Mount St. Helens on 18 May, 1980, magma had started rising from depth into the higher part of the mountain. This did change the heat flow in the edifice and caused the geyser-like eruptions of March and April 1980 (Figure 11.2; see color section). However, a more important throughput was the changing strain on the north flank of the volcano, which started bulging out to the north at nearly 1 m/day. The "unexpected response" to the increasing throughput was the catastrophic failure of the north flank, releasing the overpressure of the rocks and causing the eruption.

In 1975, I was inspired by a beautiful photograph of Old Faithful Geyser taken by Ansel Adams[1] to begin a 5-year study to explore geysers as volcanic analogs, a study now in its 29th year! In the intervening years, geysers and volcanoes were discovered in strange places in the solar system. I found myself continuing to return to Old Faithful Geyser with new questions driven by the new observations of terrestrial and non-terrestrial geysers and volcanoes. In turn, Old Faithful Geyser kept revealing more and more secrets about how these complex erupting systems work. The present chapter summarizes some of those secrets.

## 11.3  GEYSERS ON THE EARTH

Many areas on Earth that had active geysers in the past no longer have any or many geysers because of human influence. Heat from geysers is useful for direct heating of homes and greenhouses, and more complexly, for electrical generation of power. Because the heat is extracted from the ground faster than natural processes can replenish it, the natural condition of geothermal fields is always altered and diminished by this energy extraction. In Iceland—where geysers were discovered—Great Geyser is dormant, and only smaller features near it are still active. In New Zealand, much geyser heat has been diverted into power stations, and the geysers are diminished.

In Kamchatka, Russia, the "Geyser Valley" is remote and was only discovered (by a woman!) in 1941. It was largely unknown to scientists or the world until about the 1980s. Geyser Valley is a designated national preserve. It, and Yellowstone National Park, are the best-preserved geothermal areas on Earth. Yellowstone was designated a National Park by the US Congress on 1 March, 1872 because of its unique features. Thanks to the preservation of Geyser Valley and Yellowstone, the heat and fluids that power the geysers there have not been diverted or depleted by economic exploitation of the resource. Even so, pressure from tourism in Yellowstone (e.g., expansion of buildings or bridges onto geothermal ground that may host future geysers) has affected thermal features. Changes in the surficial and underground watershed, either due to natural causes, or man-made engineering efforts, also remain issues of the protection of the natural features of the Park.

Volcanic areas have two fluids that are important in eruptions: the magma inside the volcano, and hydrothermal waters around the magma (Figure 11.3). We will first

---

[1] This photograph can be found on the cover of the issue for Kieffer (1989).

look at eruptions from the hydrothermal waters—geyser eruptions—and then later look at magmatic eruptions. In the hydrothermal systems, fluids (water plus gases) are heated by magma at depth. As the water is heated, it becomes buoyant and flows upward towards the surface, where it forms hot springs, fumaroles, and geysers.

Old Faithful Geyser is the most famous geyser in Yellowstone, and it is the easiest to study because of its regularity and the open structure of its conduit. We probably know more about this geyser than any major geyser in the world. In my own studies, I have filmed it, monitored it with seismographs (Kieffer, 1984), sampled the waters for geochemical analyses (unpublished data), and with my colleagues Jim Westphal and Rick Hutchinson, put pressure–temperature sensors and eventually a small video camera into the conduit (Kieffer *et al.*, 1995; Hutchinson *et al.*, 1997). Seismicity has also been studied by Kedar (1996, 1998) and Old Faithful Geyser's response to deformation and earthquakes have been considered by Ingebritsen and Rojstaczer (1993, 1996). We do not know the 3-D geometry in detail, but it is clear that the immediate reservoir consists of interconnected fractures and caverns (Figure 11.4). The part of the conduit that has an open connection to the surface is called "the immediate reservoir"; this may connect downward through a complex plumbing system to deeper reservoirs ("ductus incognito" in Figure 11.4).

Since geyser eruptions are repeated intermittently, we can start anywhere in the cycle to describe the eruption pattern. It is common to begin with conditions at the end of an eruption. Immediately after an eruption, most or all of the water in the immediate reservoir has been discharged. Water begins to flow back into this reservoir. With time, the temperature gradually increases until the fluid becomes hot enough to boil and erupt. When I discussed the kettle on the stove earlier, I was talking about a very small volume of water that boiled when it reached the boiling temperature determined by the air pressure at the altitude of the stove. Some geysers, such as Old Faithful Geyser, are so large that the weight of the water in the immediate reservoir significantly changes the boiling pressure. Figure 11.5 is a pressure–temperature phase diagram showing how the boiling temperature increases with depth in Old Faithful Geyser (solid line), and measured temperatures and pressures in the geyser just prior to an eruption (dots). Notice that even though the water reaches 118°C in the deepest places that we could measure, it is cooler than the boiling temperature for that depth! This is a complication for which there are several possible explanations, but my favorite speculation is that there is carbon dioxide ($CO_2$) dissolved in the water. If this is true, the boiling temperature is lowered and I have speculated that the appropriate boiling curve is actually given by the measured temperatures rather than the pure boiling curve for water (Hutchinson *et al.*, 1997).

As more heat is added to the fluid over time, it changes from a liquid phase with a few steam or gas bubbles to a gas phase with entrained droplets (an aerosol). At Old Faithful Geyser, this boiling starts at the top of the water column in the immediate reservoir (Figure 11.5). The volume expansion that occurs when liquid changes to vapor causes the fluid to accelerate out of the geyser. The hottest water will form the highest part of the eruption column. Referring back to the idea of a

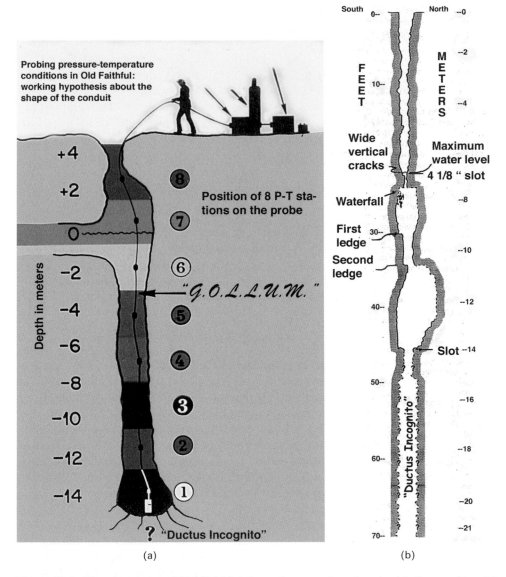

**Figure 11.4.** The upper part of Old Faithful Geyser's reservoir as imagined before probing (a) and after probing (b). It was known from earlier probe work that water never rises above about 6–7 m below the vent surface (where the 0 level is marked in (a)), but little else was known. We nicknamed the pressure–temperature probe shown here after the mythical G.O.L.L.U.M. ("Geyser Observers Lower Links to Underwater Monitoring"). We hypothesized that there was a lateral channel that drained the water at this level, but did not find evidence of it with the video camera (b). Notable features labeled are wide vertical cracks, the maximum water level, a small waterfall, two ledges, a narrow slot, and a region that we could never see ("ductus incognito"). A narrated video that shows the conduit can be viewed in low resolution at www.geyser.com, or obtained in high resolution as directed on the website.

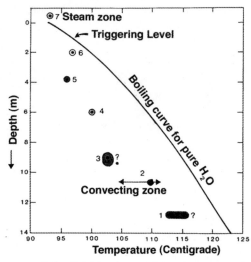

**Figure 11.5.** Temperature of the fluid in the conduit of Old Faithful Geyser vs. depth (data points corresponding to the probe stations shown in Figure 4(a)) compared to the boiling curve of pure water (solid line). Eruptions start by the ejection of water from relatively high in the conduit. Ejection of this water reduces the pressure on deeper hotter water, which then boils and erupts.

From unpublished data of J. A. Westphal, S. W. Kieffer, and R. A. Hutchinson.

complex system, the throughput that increases in Old Faithful Geyser is the addition of heat to the fluid in the conduit, and the unexpected event is the sudden change of phase of the fluid from stationary liquid with a few gas bubbles to violently erupting steam carrying water droplets out of the reservoir.

An analogy for the process is the inflation of a bicycle tire: if you are pumping air into a tire, you are increasing the throughput of gas in the tire. The tire steadily inflates as expected unless a critical value of gas throughput is exceeded, when the rubber of the tire tears and blows out. The inflating bicycle tire is a complex system. The energy stored in the compressed air inside the tire is transformed into kinetic energy as it erupts through the tear in the tire.

Using some equations for the conversion of heat energy ("enthalpy", to be accurate) to kinetic energy, I calculated that the eruption velocity of the hottest water is about 78–88 m/s, and that about 4 weight-percent of the liquid water is converted to steam (Kieffer, 1989). So, looking back at the picture of Old Faithful Geyser in Figure 11.1 (see color section), we can say that the plume from Old Faithful Geyser is composed primarily of liquid droplets, but that steam vapor is "powering" the eruption. I return below to some interesting implications of these numbers.

## 11.4   GEYSERS BECOME VOLCANOES ON MARS AND IO

If Old Faithful Geyser erupted on Mars into an atmosphere at about 0°C, the same equations used above for the parameters of an eruption of Old Faithful on Earth would produce a jet of 16 weight-percent vapor with a density of $0.03\,kg/m^3$. The maximum jet velocity at the vent would be $383\,m/s$, in contrast to the $\sim 80\,m/s$ that was derived above for Old Faithful on Earth. Using the same theory that I used to calculate the height of Old Faithful from its vent velocity, and using a slightly lower velocity of $350\,m/s$ to allow for some energy losses in the conduit, I calculate that the jet from an Old Faithful-like eruption on Mars would rise to about $485\,m$. While this is not yet a plume that we would tend to call "volcanic", it is taller than many of the "volcanic" plumes at Mount St. Helens in 1980 (Figure 11.2; see color section).

Abundant photographic evidence suggests the presence of both volcanoes and liquid on Mars in the past. Water still exists at present in the frozen polar caps, and probably as subsurface ice widely distributed around the planet. As discussed in Chapters 6, 7, and 10, there is ample evidence for volcanism on Mars. Recent high-resolution images from the Mars Orbiter Camera (MOC) and Mars Orbiter Laser Altimetry (MOLA) data show detailed evidence for magma–ice interactions. Most scientists believe that the fluid involved in Martian volcanism was $H_2O$, but the role of $CO_2$ is not yet well-understood.

Unfortunately, although there are volcanic landforms on Mars, there is no current active volcanic activity and we can only speculate through computer simulations about the nature of the volcanism (Kieffer, 1995). Fortunately, there are erupting volcanoes elsewhere in the solar system.

In 1979, the Voyager 1 spacecraft was approaching Jupiter, and its cameras captured some remarkable images of the satellite Io (Figure 11.6; see color section). Gene Shoemaker called me from the Jet Propulsion Laboratory (JPL) where teams were working on data being transmitted back from the Voyager 1 spacecraft approaching Jupiter to tell me—in a very excited voice—that Voyager 1 cameras had captured huge (70–280 km) plumes erupting above the surface of Io! Chapter 8 reviews these observations as well as more recent data. Gene asked me if I could apply my geyser theories to the analysis of these plumes. "Oh, dear me, I thought, I am only 3 years into my plan to study geysers as volcanic analogs and now I have to think about some of the weirdest geysers imaginable!"

The planet was covered with sulfur and sulfur dioxide, and the first hypotheses were that at least some of the plumes were made of sulfur dioxide emerging from a shallow reservoir in the crust of Io (Figure 11.7). Later observations (Chapter 8) showed higher temperature sulfur plumes and very hot silicate lava lakes. The heat to melt rocks to produce the lavas, sulfur, and sulfur dioxide is created by tidal friction as Jupiter pulls Io one way, and three other satellites (Ganymede, Calisto, and Europe) pull other ways. In the sense of a complex system, in some way not yet well understood, the tidal strain is probably the throughput, and the eruptions are the unexpected events.

The challenges in modeling Io's plumes were many: initial observations pointed to relatively cold plumes (e.g., only slightly over $100°C$), although later Near-Infrared Mapping Spectrometer (NIMS) observations have revealed some of the hottest known volcanic eruptions in the solar system (Chapter 8). The plumes were erupting into a near vacuum (atmospheric pressure $\sim 10^{-9}$–$10^{-12}$ bar), gravity is only 0.18 of that on Earth, and the spacecraft data tell us that the erupting fluids

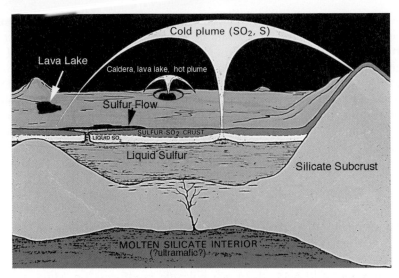

**Figure 11.7.** A schematic diagram of the proposed structure of the upper crust on Io. The scale is tens to hundreds of kilometers vertically, and hundreds of kilometers horizontally. The molten silicate interior may have differentiated to form a silicate crust that forms high mountains on Io. The upper crust is continuously covered with recycled S, $SO_2$, and possibly silicate ash, from the vigorous volcanism. Due to the thermal gradient, these components are believed to separate into zones of molten sulfur, liquid $SO_2$ and solid S and $SO_2$ ices that lie above the silicates.

are sulfur-based, not water-based. I needed to develop a completely new set of phase diagrams for these fluids in order to calculate the enthalpy change as the fluid is erupted and I had to deal with the difficult problem of phase changes during the eruption (Smith *et al.*, 1979). From the enthalpy changes I was able to show that even a relatively cold (100°C) sulfur dioxide geyser had enough energy to rise to the observed heights of hundreds of kilometers. Additional enthalpy is available in the hotter plumes and thus a wide variety of plume source conditions and eruption styles are possible (Kieffer, 1982a).

The final challenge was posed by the high-pressure nature of the material erupting into a low-pressure environment: very high-pressure gradients produce very high-speed flows and shock waves. We have very little intuition about such flows because we live in a world of fairly constant pressure. Perhaps the highest pressure ratio we generally encounter is the blowout of a bicycle or automobile tire. Our world is a "subsonic" world. However, under certain conditions, a world can become dominated by "supersonic" flow, and Io is one such place. These worlds are described by the ratio of the flow velocity ($u$) to the fluid sound speed ($c$). This ratio is called the Mach number ($M = u/c$). When $M < 1$, the flow is subsonic, and when $M > 1$, the flow is supersonic. Our main encounter with the word supersonic, is in the context of airplanes: when they fly faster than the speed of sound in air ($\sim 330\,\text{m/s}$), they are supersonic. Shock waves create audible booms when such aircraft pass overhead.

However, in the context of volcanic processes, we care primarily about the sound speed of the erupting fluid, and this can be surprisingly low. For example, the sound speed in boiling water in a geyser reservoir can be as low as a few meters per second—a speed that could allow a good sprinter running through boiling water to be supersonic! With this in mind, I returned to my project of understanding of Old Faithful Geyser, wondering if perhaps its fluid was flowing supersonically. I calculated the sound speed for water containing 4% vapor (derived above), and found that it was 88 m/s. Is this just coincidentally very close to the derived vent velocity of 78–88 m/s? I think not! I conclude that Old Faithful Geyser probably reaches $M \sim 1$ at about 7 m depth, where the conduit is constricted down to about 0.1 m (Figure 11.4(b)). This would be such a weak transition to sonic or supersonic flow that it probably decelerates to subsonic conditions immediately above this point and no shocks are produced.

However, conditions are different on Io, where plumes erupt into a near-vacuum. The pressure gradient between the vent and atmosphere is possibly more than ten orders of magnitude, producing very high velocities in a very cold gas (sound speed decreases with decreasing temperature).

We (Smythe et al. 2001) used a computer model to simulate eruptions from 1 bar vent pressure (a pressure that must be attained at or slightly below the surface) into a tenuous atmosphere of $10^{-7}$ bar (Figure 11.8(top)). As the plume evolves, material accelerates outwards at supersonic speeds. A shock wave develops around the edge of the expanding plume (dark band).

The continuum model used above breaks down at very low pressures because the distance between gas molecules becomes significant. New work (Zhang et al., 2004) using a direct simulation Monte Carlo (DSMC) method addresses the far field aspects of the plume (Figure 11.8(bottom)). In this model, the "vent" pressure is tens of nanobars and the vent is taken to be much larger than the vent for the continuum model (i.e., the vent is a "virtual vent" above the real physical vent). The results of the two models are qualitatively similar and the DSMC model can be directly compared to Figure 11.6(b) (see color section). Further research connecting the two approaches should lead to a deeper understanding of the relation between the rarified plume and volcanic conduit conditions.

Compared to Old Faithful Geyser or large Plinian eruption columns on Earth, the plumes on Io have a very unusual shape. The Plinian plumes on Earth typically rise straight above the volcano with a gently expanding diameter (Figure 11.9). The inner sheath of fluid ascending from Old Faithful Geyser also rises vertically (Figure 11.1; see color section), as did the fluid erupting from the early eruptions at Mount St. Helens (Figure 11.2, left edge of the eruption column; see color section). These jets are referred to as "pressure balanced" because the pressure in the jet is the same as that of the surrounding atmosphere.[2] On Io, both the continuum and DSMC models show that the fluid blasts out horizontally as well as vertically. This expansion is driven by the very large pressure gradient between the vent and the near-vacuum atmosphere. That is, the flow is internally supersonic and

---

[2] Jets can be pressure balanced and supersonic, but for simplicity we ignore that case here.

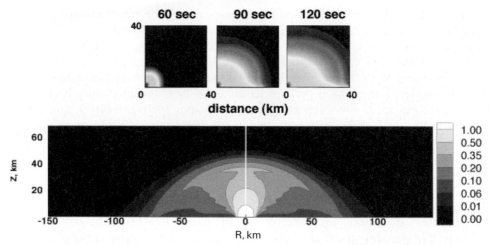

**Figure 11.8.** Computer simulation of Prometheus-type plumes on Io. (Top) A continuum model in which the vent pressure is 1 bar, and atmospheric pressure is $10^{-7}$ bar. Micron-sized particles are embedded in $SO_2$ gas. The parameter plotted is the log of the volume fraction of particle concentration. The particles are 1 weight-percent of the plume. The scale is smaller than Prometheus in Figure 11.6(b) because of the limitations of the computer, but the initial conditions are similar to those believed to exist in the reservoir at Prometheus. Higher particle concentrations are dark near the center, and lower concentrations are light. The tenuous atmosphere is also dark (upper right). The alternating light–dark–light near the top of the plumes indicates a shock wave structure, a so-called "canopy shock". (Bottom) a DSMC simulation of the behavior of 1 nm particles in a gas plume. In this simulation, the vent pressure is in nbar ($10^{-9}$ bar), and atmospheric pressure is approximately five orders of magnitude lower. The parameter plotted is dust column density.
(Top) From unpublished data of S.W. Kieffer, R. Lopes, and W. Smythe. (bottom) From Zhang *et al.* (2004)


is termed "overpressured" because of the high pressure in the vent relative to the atmosphere.

If the pressure in the plume is high compared to the atmosphere (e.g., a few hundred to thousands times atmospheric), but not as high as in the plumes on Io (billions of times atmospheric), then the flow expands out laterally to a shape determined by the pressure ratio. This case is nicely illustrated by typical conditions in discharging geothermal bores (Figure 11.10).

I was struggling to learn these supersonic flow concepts to apply to the Ionian volcanoes and geysers when Mount St. Helens came to life in 1980, and little did I know that they would be relevant to the disasterous eruption of 18 May, 1980.


## 11.5    FROM IO TO MOUNT ST. HELENS

Mount St. Helens started erupting on 27 March, 1980. I first heard of these eruptions when a volcanologist visiting the US Geological Survey in Flagstaff, Arizona,

**Figure 11.9.** Photo of Mt. St. Helens, 18 May, 1980. (Compare with Figures 11.1 and 11.2 (see color section) to see similarities in the vertical nature of the upgoing part of the plume. (The plume is $\sim 1$ km diameter at the base.)
Austin Post, US Geological Survey.

**Figure 11.10.** A discharging geothermal bore in Iceland. Note the lateral flaring at the exit plane of the bore. This indicates an overpressured jet. This discharge roared like a supersonic engine, suggesting that the flow is internally supersonic. The noise would originate mainly from a shock wave that is probably 3–4 bore diameters up into the plume, not visible because of the sheath of steam and entrained air surrounding it. I tried to throw a rock into this plume to get a feel for flow velocities. I was ever so suprised when the rock bounced back towards me, unable to penetrate into the interior of the high-pressure discharge.
Photo by S. Kieffer.

mentioned this new eruption. I could not resist this unique chance to extend my emerging geyser ideas to eruptions of this size, and so I grabbed my Super-8 movie camera, and hopped on a plane to Mount St. Helens (oh, would I have loved to have had modern digital videos!). The eruptions in late March and April were very much like geyser eruptions, both in scale (hundreds of meters) and frequency (every few hours). They appeared much more "volcanic" than geyser-like because the erupting steam was carrying a heavy load of rock and ash from the enlarging crater on the summit of the mountain, but other than this, the eruptions resembled huge geyser eruptions (Figure 11.2; see color section). I filmed and timed these eruptions to try to figure out the heat and water budget of the mountain, and produced the basic sketch shown in Figure 11.11(a) of my ideas about what was happening in the mountain. The "heat budget" inside the mountain had been increasing. In the context of complex systems discussed above, the increasing throughput of heat resulted in the unexpected: the geyser-like eruptions. Possibly, as indicated in Figure 11.11(a) magma had shifted in the mountain, causing a change in the heat budget.

In late April, these geyser-like eruptions stopped. It is likely that the throughput—the heat—had diminished to the point where it could no longer melt the snow and glaciers to power the geyser-like eruptions. However, it had simultaneously become clear to all working there that the mountain was changing. The north flank was discovered to be bulging out at a strain rate of nearly 1 m/day, and it was clear that magma moving higher into the mountain was causing the bulging. The critical throughput changed from heat to strain. Danger zones were defined and the public was forbidden access to a fairly large area around the mountain.

On 18 May, 1980, the strain throughput changed the behavior of the mountain dramatically, causing the complete failure of the north flank into a large complex landslide. The overburden pressure on hot fluids and hot rocks was reduced and they erupted out toward the north sector in an event that became known as "the lateral blast".[3] The lateral blast devastated an area of over $500 \, km^2$ removing trees and soil near the source (Figure 11.12(a)), snapping trees off of their rootballs and carrying them away (Figure 12(b)), and toppling trees at larger distances from the blast (Figure 12(c)). At even further distances, trees remained standing, but were singed by the hot gases in the eruption cloud. Devastation reached out to nearly 30 km to the north.

Remnants of trees and downed trees were good indicators of the direction of the blast (Figures 11.12(b, c)). I carefully mapped the directions of downed trees, and then generalized the devastated area into the three parts shown in Figure 11.13—"direct", "channelized", and "singed" blast zones. Note the similarity in shape of the direct blast zone near the source to the geothermal plume near the base in Figure 11.10.

It was impossible to do a fully 3-D flow analysis for the conditions of the blast and the rugged topography over which it traveled. In fact, as this book is published, such an analysis has still not been done 24 years later. However, by simplifying the

---

[3] The lateral blast should not be confused with the Plinian eruption that developed about 4 h later (Figure 11.9)

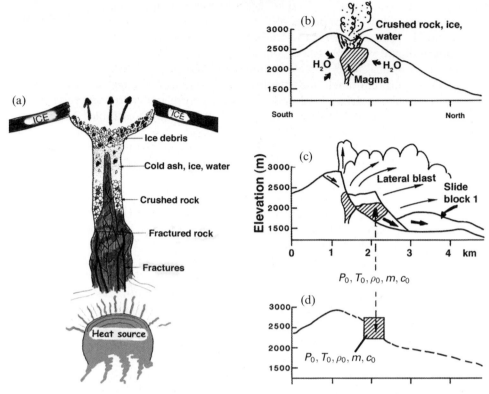

**Figure 11.11.** A schematic diagram of the conditions in Mount St. Helens in March through May 1980. (a,b) During late March, April, and into May, magma moved in the mountain, eventually causing a bulge on the north flank. This is a sketch that I made in the field in April of 1980 to illustrate to the press and public the conditions in the mountain. (c) The increasing strain on the north flank from the magma movement eventually caused the flank to collapse catastrophically in a series of landslides. This released the pressure on the inner parts of the volcano, allowing gases to expand to start the lateral blast. (d) A schematic diagram showing how I simplified this complex process for modeling. I assumed that there was a reservoir with average initial properties (pressure $P_0$, temperature $T_0$, density $\rho_0$, mass ratio of solids to gas $m$, and sound speed $c_0$).

thermodynamic properties of the erupting fluid, and making simplifying assumptions about the terrain, I could use some of the rocket nozzle theory that I had begun to explore to explain volcanic eruption processes on Io to give a semiquantitative model of the fluid dynamics within the blast.

There are two especially important parameters that determine rocket nozzle behavior. We have discussed one, the pressure ratio between the fluid and the atmosphere at the exit plane (for a volcano, we call this the vent). At Mount St. Helens, I estimated this ratio to be close to 150 : 1 (125 bar pressure at the vent divided by 0.87 bar, the atmospheric pressure at the appropriate elevation at Mount St. Helens). At this pressure ratio, gravity is not an important term in the momentum equation in

**Figure 11.12.** Damage from the 18 May, 1980 lateral blast at Mount St. Helens. (a) A piece of timbering equipment located in the direct blast zone at Mount St. Helens, showing the nature of the material being carried in the lateral blast. In this zone, all trees, along with their root balls and some soil and underlying volcanic deposits were stripped from the landscape. (b) A destroyed tree stump, with shredded fragments pointing toward downstream in the blast. (c) Downed trees in the channelized blast zone, showing the general direction of flow (right to left). (a) and (b) are several meters in width, (c) is several hundred meters.
Photos by S. Kieffer.

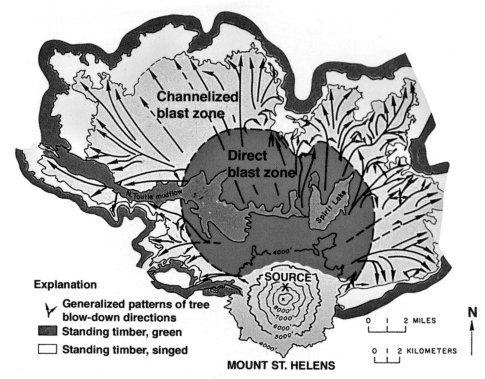

**Figure 11.13.** A map summarizing the devastation caused by the lateral blast at Mount St. Helens. I used detailed studies of the directions of the downed trees to create this map. The "X" on the north flank marks the general center of the region that failed into the landslide that triggered the blast. In the inner direct blast zone, trees were felled nearly radially around the vent, independent of huge valleys and ridges. In the outer channelized blast zone, the tree blow-down direction indicated that the blast was channeled into valleys by ridges blocking its path. In the singed zone trees remained standing but were singed by the hot gases of the blast passing overhead. The singed zone is surrounded by standing green trees.
From Kieffer (1981) and Kieffer (1982b).

the near field. The flow expanded supersonically out of the vent. This means that we need to estimate the sound speed at the vent $c_0$. The fluid erupting into the lateral blast was a complicated mixture of vapor (mostly steam), solids ranging in size from less than one millimeter to greater than several meters, organics torn up from the forest, and ice from the glaciers. The concept of sound speed in such a heterogeneous material is ill-defined, but it is certainly low. Compared to steam with a sound speed of 440 m/s, the sound speed is reduced by the entrainment of all of the particulate ash and dirt. For a simple analogy, and working model, the sound speed is comparable to that of a heavy molecular weight gas. For the lateral blast, I estimated that the sound speed was about 100 m/s (Kieffer, 1981, 1982b). Perhaps by coincidence, or not, the head of the lateral blast traveled at a fairly constant velocity of 100 m/s across much of the devastated area. This velocity is about 1/3 of the speed of sound in air, and thus no "sonic booms" were heard near the volcano (Reed, 1987).[4]

For the model, I assumed that the vent velocity was 100 m/s (i.e., $M = 1$), and then I solved the compressible flow equations to examine the steady-state flow conditions inside the blast. Model results are shown in Figure 11.14.

Behind the flow head, conditions are different from those within the flow head. Because of the huge overpressure, the fluid expands laterally (Figure 11.14(a)), reducing the pressure. In Figure 11.14(a), the second ray of numbers shows the ratio of the pressure at a given contour to the initial vent pressure, 125 bar. Atmospheric pressure at the average elevation of the vent was 0.87 bar. Thus, at the contour 0.0066, the pressure is reduced to atmospheric pressure. However, the expansion is non-linear, the system "overshoots", and the pressure keeps decreasing. This produces a zone, shown by the stippled pattern, in which the pressure inside the blast is actually less than atmospheric, a moderate vacuum condition. This is certainly non-intuitive! This overshooting continues until shock waves occur around the expanding fluid; these shock waves increase the pressures back towards atmospheric. The shock waves at the sides of the flow are called "intercepting" or "barrel shocks", and a very strong shock in the front of the flow is called the "Mach disk shock".

A second non-intuitive result is that the velocity keeps increasing with expansion in supersonic flow. This is shown in Figure 11.14(a) by the first ray labeled $M$, the Mach number. The Mach number increases from 1 at the vent to nearly 3.5 just before the Mach disk shock. Thus, even though the flow front—which is somewhere downstream of the shocks in this model—is traveling only at 100 m/s, the fluid behind the flow front and inside of the blast where we can't see it is traveling at about 300–350 m/s! It is supersonic, as indicated by $M > 1$ in Figure 11.14(a). This velocity is decreased toward the flow head velocity by the Mach disk shock; entrainment of atmospheric air further slows the material after it passes through the shock.

When scaled to the vent size of the lateral blast ($\sim$1 km diameter where the north flank failed) and superimposed on the map of the tree devastation of Figure 11.13, the flow map of Figure 11.14(a) shows a rather remarkable correlation with the shape of the devastated area and its zonation (Figure 11.14(b)). The flaring of the blast to the east and west of the northerly vent direction is caused by the pressure gradients and is directly comparable to that of erupting geothermal fluid shown in Figure 11.10. The direct blast zone coincides with the supersonic flow region, and the channelized blast zone corresponds to the subsonic region beyond the shocks. The shock waves serve two major purposes: they slow the flow down, and they bring it from subatmospheric pressures back to higher pressures. Until we have much better computational codes we cannot quantify this process, but I believe that beyond the shock waves the pressure gradients are small enough that gravity now becomes an important term in the momentum equation. In this case, the flow becomes subsonic,

---

[4] Booms were heard as far north as Vancouver, British Columbia, but these were caused by reflections and coalescences of the pressure waves from the blast (Kieffer, 1982b) into audible booms (Reed, 1987).

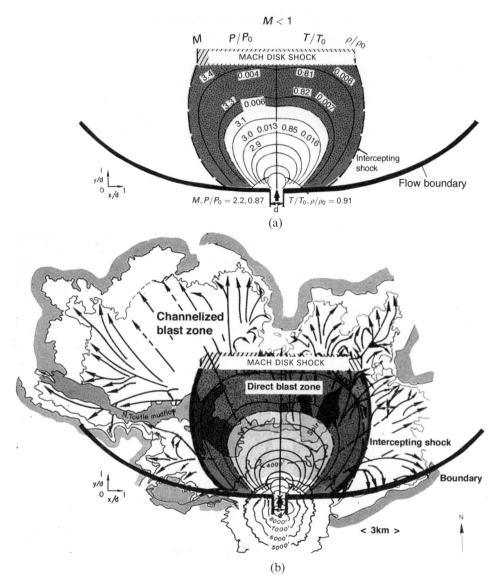

**Figure 11.14.** (a) Model results for contours of constant properties in the lateral blast. (b) The model calculations from (a) scaled to the dimensions of Mount St. Helens and superposed on the map of the devastated zones from Figure 11.13.
From Kieffer (1982) and Kieffer and Sturtevant (1984).

as indicated by $M < 1$ in Figure 11.14(a). Thus, the flow is divided into two regions: an inner core of high-velocity supersonic flow with complicated pressure gradients and where gravity is not important compared to these pressure gradients, and an outer zone of lower velocity, subsonic flow in which normal gravitational effects are important. The flow only started following the terrain in this "channelized blast zone".

Although we had no direct measurements within the lateral blast, some of these conclusions can be tested. There was a unique window of time in the 1980s when the erosion surface carved by the blast allowed features to be mapped in the erosion surface produced by the blast. Erosion by the blast scoured the pre-existing forest and substrate. Large furrows of the scale of meters in width and hundreds of meters in length were found eroded into the substrate in a number of places, and from these features it was possible to infer that velocities significantly greater than the flow front had indeed caused scouring of the land (Kieffer and Sturtevant, 1984).

This lateral blast of 1980 was an unusual, but not unique, event in modern volcanology. Unfortunately, after 23 years, most of the field evidence has disappeared to erosion or vegetation. Much work remains to truly understand the dynamics and power of these blasts.

## 11.6   FROM MOUNT ST. HELENS TO TRITON AND TITAN

With the work at Mount St. Helens, I had come prematurely to a volcanological application of my unfinished geyser work, and I was happy to get back to working on the problem of Old Faithful in the mid-1980s, designing a pressure–temperature probe that we could put into Old Faithful and later, an ice-cooled camera to document the geometry of the conduit (Figure 11.4(b)). I was totally unprepared for the next phone call from Gene Shoemaker in 1989: "Sue, Voyager 2 has discovered plumes on Triton (a satellite of Neptune)! They are erupting from an icy nitrogen and methane surface that is only at 38 K ($-235°C$)! Can you apply your geyser theory to this problem to see if these could be cold geysers?" Uh-oh, I thought, here we go again!

Triton is the largest satellite of Neptune. It has a diameter of 2,700 km, about 80% of the diameter of our own moon. In 1989, the cameras on Voyager observed 4 plumes erupting in the southern hemisphere; one is shown in Figure 11.15 (see also Chapter 9). The plumes had diameters between tens of meters and 1.5 km. They rose to an altitude of 8 km, where winds apparently can carry the plume material at least 100 km to the west. In the photos of the plumes, dark material (which may be carbonaceous material, silicates, or metals) and light material also descend towards the surface. The light material might be snowflakes of nitrogen ice condensing as vapor is carried downwind from the sites of the plumes. There are also at least 100 dark streaks on the surface near these plumes that may be deposits from previously active plumes, and some of these are 150 km long.

In order to form plumes on Triton, the weak solar energy at the orbit of Neptune must be captured. There are several suggestions for this mechanism, the most prominent being "nitrogen geysers" (Soderblom *et al.*, 1990), and "dust devils" (Ingersol and Tryka, 1990). Kirk *et al.* (1995) review these models and heat sources. In the sense of complex systems, the increasing throughput appears to be the seasonal change in solar radiation as Neptune orbits the sun (Soderblom *et al.*,

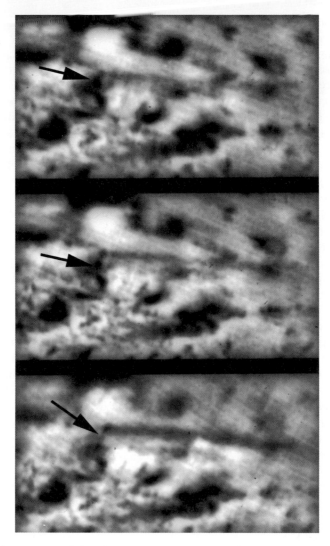

**Figure 11.15.** A sequence of three images from Voyager 2 of Triton. These images have been projected onto a spherical surface and have been spatially filtered to enhance detail so that the viewer can see the plumes in different projections as the spacecraft flew over the planet. The plumes rise into small, dark, dense clouds at their tops, and produce long, horizontal, diffuse dark clouds that trail off to the west (right) in the atmospheric winds, best seen in the bottom image.
NASA.

1990), and the unexpected events are these eruptions. I was working on the hypothesis that these were nitrogen plumes, and in order to try to understand these plumes, I had to decide whether or not they were likely to be supersonic. Triton's surface pressure is about 14 μbar ($14 \times 10^{-6}$ bar; about 1/70,000 that of Earth's surface pressure). This pressure is consistent with the vapor pressure equilibrium of nitrogen gas over nitrogen ice at 38 K. Because there is very little energy available to substantially increase the temperature of the ice to form geysers, it is also unlikely that the geysers will be driven by large pressure gradients, and so we examined conditions for subsonic geysers (i.e., "jets" driven by small energy changes converted into momentum).

Investigators had noted that sunlight penetrates roughly a meter through the nitrogen ice cap on Triton, where it probably encounters dark organic polymers derived from methane. These dark polymers would allow solar heating, which would warm the nitrogen enough to vaporize it locally (Kirk *et al.*, 1990). If it migrated to cracks, it could erupt explosively into the observed plumes. Calculations of the enthalpy available show that if the nitrogen were heated only 4°C above its surface temperature to −231°C, a velocity of 180 m/s could be obtained, and this could drive the plumes to the observed height of 8 km (Soderblom *et al.*, 1990).

Finally, geysers of methane have now been postulated to be possible on Titan, the large and icy satellite of Saturn (Lorenz and Mitton, 2002; Figure 11.16, color section). The surface temperature on Titan is −178°C, and it has a thick atmosphere of nitrogen and methane with small amounts of oxygen. It is possible that lakes of liquid ethane underlain by frozen methane, ammonia, and water exist near the surface. Lorenz (2002) proposed that if areas of concentrated heat flux exist on Titan, geysers of ethane–methane–nitrogen are possible. Using concepts similar to those described here and considering Old Faithful Geyser, Io, and Triton, Lorenz shows that geysers are possible, that eruption velocities would be about 25 m/s, and that the periodicity and structure of the geysers should be similar to those on Earth! In terms of complex systems theory, the increasing throughput could be either gravitational heating, thermal tides in the atmosphere, or solar heating as for Triton, and the unexpected events would be the geyser eruptions.

The Huygens probe is due to separate from the Cassini spacecraft on 25 December, 2005, and to send back 1,100 images on 14 January, 2005. Another volcano world??

## 11.7  SUMMARY

In this chapter, I have tried to show you how geysers and volcanoes in the solar system show a wide range of behaviors, with emphasis on their eruptions as "unexpected events" in the context of modern complex systems theory. Many problems remain for volcanologists of the future. We know from our general geologic observations that these systems are incredibly complicated, and that "complicated" has a different meaning than "complex"! Complications include: geometry of conduits and vents; thermodynamic properties of the erupting fluids; stresses and strains surrounding the volcanoes; and atmospheric or suboceanic conditions. Complex system theory tells us that prediction of the timing and nature of the "unexpected events" will not be an easy task, and so much remains to challenge volcanologists.

## 11.8  EPILOGUE

I began this chapter with musings on a walk through Yellowstone National Park, wondering at the sounds of geysers erupting on a winter's night, and thinking of how eruptions into the near vacuum of Io would make geysers difficult to distinguish from volcanoes on that planet. Indeed, our definition of the word "geyser" has been magnificently broadened by the discoveries of the space program. Similarly, as the chapters of this book have demonstrated, the word "volcano" has taken on ever-broader meanings as we have explored the oceans and the planets in addition to our ground-based home. Now, as I begin my 30th year of thinking and wandering around Yellowstone again, and as you begin thinking about volcanoes, those same musings will extend to the wonderment of what may be revealed in the volcano worlds of the future.

## 11.9  REFERENCES

Hutchinson, R.A., Westphal, J.A., and Kieffer, S.W. (1997) In situ observations of Old Faithful Geyser. *Geology*, **25**(10), 875–878.

Ingebritsen, S.E., and Rojstaczer, S.A. (1993) Controls on geyser periodicity. *Science*, **262**, 889–892.

Ingebritsen, S.E., and Rojstaczer, S.A. (1996) Geyser periodicity and the response of geysers to deformation. *J. Geophys. Res.*, **101**, 21891–21906.

Ingersoll, A.P. and Tryka, K.A. (1990) Triton's plumes: The dust devil hypothesis. *Science*, **250**, 435–439.

Kedar, S., Kanimou, H., Stürtevant, B. (1996) The origin of harmonic tremor at Old Faithful Geyser. *Nature*, **379**, 708–711.

Kedar, S., Kanimou, H., Stürtevant, B. (1998) Bubble collapse as the source of tremor at Old Faithful Geyser. *J. Geophys. Res.*, **89**, 8253–8268.

Kieffer, S.W. (1981) Blast dynamics at Mount St. Helens on 18 May 1980. *Nature*, **291**, 568–570.

Kieffer, S.W. (1982a) Dynamics and thermodynamics of volcanic eruptions: Implications for the plumes on Io. In: D. Morrison (ed.), *Satellites of Jupiter*. University of Arizona Press, Tucson, AZ, pp. 647–723.

Kieffer, S.W. (1982b) Fluid Dynamics of the May 18 Blast at Mount St. Helens (Professional Paper 1950). US Geological Survey, Reston, VA.

Kieffer, S.W. (1984) Seismicity at Old Faithful Geyser: An isolated source of geothermal noise and possible analogue of volcanic seismicity. *J. Volcanology and Geothermal Res.*, **22**, 59–95.

Kieffer, S.W. (1989) Geologic Nozzles. *Reviews of Geophysics*, **27**(1) 3–38.

Kieffer, S.W. (1995) Numerical models of caldera-scale volcanism on Earth, Venus, and Mars, *Science*, **269**, 1385–1391.

Kieffer, S.W. and Sturtevant, B. (1984) Laboratory studies of volcanic jets. *J. Geophys. Res.*, **89**, 8253–8268.

Kieffer, S.W., Gautier, R.L., McEwen, A., Smythe, W., Keszthelyi, L., and Carlson, R. (2000) Prometheus: Io's wandering plume. *Science*, **288**, 1204–1208.

Kieffer, S.W., Westphal, J.A., and Hutchinson, R.A. (1995) A journey toward the center of the Earth: Video adventures in the Old Faithful conduit. *Yellowstone Science*, **3**(3), 2–5.

Kirk, R.L., Brown, R.H., and Soderblom, L.A. (1990) Subsurface energy storage and transport for solar-powered geysers on Triton. *Science*, **250**, 424–429.

Kirk, R.L., Soderblom, L.A., Brown, R.H., Kieffer, S.W., and Kargel, J.S. (1995) Triton's plumes: Discovery, characteristics, and models. In: D.P. Gruikshank (ed.) *Neptune and Triton*. University of Arizona Press, Tucson, AZ, pp. 949–989.

Lorenz, R.D. (2002) Thermodynamics of geysers: Application to Titan. *Icarus*, **156**, 176–183.

Lorenz, R. and Mitton, J. (2002) *Lifting Titan's Veil*. Cambridge University Press, Cambridge, U.K., 268 pp.

Reed, J.W. (1987) Air pressure waves from Mount St. Helens eruptions. *J. Geophys. Res.*, **92**, 11979–11992.

Smith, B.A., Shoemaker, E.M., Kieffer, S.W., and Cook II, A.F. (1979) The role of SO$_2$ in volcanism on Io. *Nature*, **280**, 738–743.

Smythe, W.D., Kieffer, S.W., and Lopez-Gautier, R. (2001) Plume models and pyroclastic flows on Io. *Lunar and Planet. Sc. XXXII*, Abstract #2129.

Soderblom, L.A., Kieffer, S.W., Becker, T.L., Brown, R.H., Cook II, A.F., Hansen, C.J., Johnson, T.V., Kirk, R.L., and Shoemaker, E.M. (1990) Triton's geyser-like plumes: Discovery and basic characterization. *Science*, **250**, 410–415.

Strom, R.G. and Schneider, N.M. (1982) Volcanic eruption plumes on Io. In: D. Morrison (ed.) *Satellites of Jupiter*. University of Arizona Press, Tucson, AZ, pp. 598–633.

Zhang, J., Goldstein, D.B., Varghese, P.L., Trafton, L.M., Moore, C., and Miki, K. (2004) Numerical modeling of Ionian volcanic plumes with entrained particulates. *Lunar and Planet. Sc. XXXV*, Abstract #1972.

# Index

Printing: Mercedes-Druck, Berlin
Binding: Stein+Lehmann, Berlin